Field Guide to Common Marine Fishes and Invertebrates of Alaska

Susan C. Byersdorfer
and Leslie J. Watson

Published by Alaska Sea Grant College Program
University of Alaska Fairbanks

Sea Grant Alaska

$35
SG-ED-67

Elmer E. Rasmuson Library Cataloging in Publication Data:

Byersdorfer, Susan C.

Field guide to common marine fishes and invertebrates of Alaska / by Susan C. Byersdorfer and Leslie J. Watson.—Fairbanks, Alaska : Alaska Sea Grant College Program, University of Alaska Fairbanks, 2010.

Xviii, 358 p. : ill., map—(Alaska Sea Grant College Program, University of Alaska Fairbanks; SG-ED-67)

Includes bibliographical references.

1. Marine fishes—Alaska—Identification. 2. Marine invertebrates—Alaska—Identification. I. Title. II. Watson, Leslie J. III. Series: Alaska Sea Grant College Program, University of Alaska Fairbanks; SG-ED-67.

QL623.2.B94 2010

ISBN 978-1-56612-149-1

Second Printing

Credits

This book is published by the Alaska Sea Grant College Program, supported by the U.S. Department of Commerce, NOAA National Sea Grant Office, grant NA06OAR4170013, project A/161-01; and by the University of Alaska Fairbanks with state funds. The University of Alaska is an affirmative action/equal opportunity employer and educational institution. Support for publishing cost was also provided by the Alaska Department of Fish and Game, Division of Commercial Fisheries, Alaska Fisheries Information Network.

Sea Grant is a unique partnership with public and private sectors combining research, education, and technology transfer for public service. This national network of universities meets changing environmental and economic needs of people in our coastal, ocean, and Great Lakes regions.

Book design and layout by Janelle Peterson of Kodiak, Alaska, and Jen Gunderson of Alaska Sea Grant; cover design by Jen Gunderson; project coordination by Kurt Byers; and project management and copyediting by Sue Keller, of Alaska Sea Grant.

Front cover: the predaceous aeolis, *Flabellina triophina*, glides over a red alga on a rock substrate 9 meters deep off Atka Island in the central Aleutian Islands. Photo by Stephen Jewett. Dr. Jewett is a research professor and dive safety officer at the University of Alaska Fairbanks School of Fisheries and Ocean Sciences. He is one of a pioneering few to extensively explore the Aleutian Islands by scuba.

Back cover: Héloïse Chenelot, research technician at the University of Alaska Fairbanks Institute of Marine Science, took the photo of a female rock greenling (*Hexagrammos lagocephalus*) near Amchitka Island, Alaska, at a depth of 10 m. Also visible are orange sponges.

Alaska Sea Grant College Program
University of Alaska Fairbanks
P.O. Box 755040
Fairbanks, Alaska 99775-5040
Toll free (888) 789-0090
(907) 474-6707 • fax (907) 474-6285
www.alaskaseagrant.org

Printed in China

TABLE OF CONTENTS

ACKNOWLEDGMENTS .... **xvii**

PROJECT SPONSORSHIP .... **xvii**

ABOUT THE AUTHORS .... **xviii**

INTRODUCTION **1**

HOW TO USE THIS GUIDE **3**

LAMPREY **4**

Arctic lamprey *Lethenteron camtschaticum*

SHARKS **4**

Pacific sleeper shark *Somniosus pacificus*
Spiny dogfish *Squalus acanthias*

SKATES **7**

External Features of Skates **7**

Longnose skate *Raja rhina*
Big skate *Raja binoculata*
Alaska skate *Bathyraja parmifera*
"Leopard" skate *Bathyraja* sp. cf. *parmifera*
Mud skate *Bathyraja taranetzi*
Aleutian skate *Bathyraja aleutica*
Bering skate *Bathyraja interrupta*
Commander skate *Bathyraja lindbergi*
Whiteblotched skate *Bathyraja maculata*

External Features of Soft-Rayed Fishes **16**

TROUTS **16**

Rainbow trout *Oncorhynchus mykiss*
Cutthroat trout *Oncorhynchus clarkii*
Dolly Varden *Salvelinus malma*

SALMON **18**

Pink (humpback) salmon *Oncorhynchus gorbuscha*
Coho (silver) salmon *Oncorhynchus kisutch*
Chinook (king) salmon *Oncorhynchus tshawytscha*
Chum (dog) salmon *Oncorhynchus keta*
Sockeye (red) salmon *Oncorhynchus nerka*

HERRING **22**

Pacific herring *Clupea pallasii*

SMELTS 22

Rainbow smelt *Osmerus mordax*
Capelin *Mallotus villosus*
Eulachon *Thaleichthys pacificus*

CODS 25

Pacific (gray) cod *Gadus macrocephalus*
Pacific tomcod *Microgadus proximus*
Saffron cod smelt *Osmerus mordax*
Capelin *Mallotus villosus*
Saffron cod *Eleginus gracilis*
Walleye pollock *Theragra chalcogramma*

ROCKFISHES 27
External Features of Rockfishes 27

Shortspine thornyhead *Sebastolobus alascanus*
Yelloweye rockfish *Sebastes ruberrimus*
Shortraker rockfish *Sebastes borealis*
Roughcye rockfish *Sebastes aleutianus*
Blackspotted rockfish *Sebastes melanostictus*
Pacific ocean perch *Sebastes alutus*
Harlequin rockfish *Sebastes variegatus*
Sharpchin rockfish *Sebastes zacentrus*
Northern rockfish *Sebastes polyspinis*
Canary rockfish *Sebastes pinniger*
Tiger rockfish *Sebastes nigrocinctus*
Redbanded rockfish *Sebastes babcocki*
China rockfish *Sebastes nebulosus*
Black rockfish *Sebastes melanops*
Dark rockfish *Sebastes ciliatus*
Dusky rockfish *Sebastes variabilis*
Silvergray rockfish *Sebastes brevispinis*
Yellowtail rockfish *Sebastes flavidus*
Quillback rockfish *Sebastes maliger*

SABLEFISH 44

Sablefish (blackcod) *Anoplopoma fimbria*

GREENLINGS 44

Atka mackerel *Pleurogrammus monopterygius*
Lingcod *Ophiodon elongatus*
Kelp greenling *Hexagrammos decagrammus*
Rock greenling *Hexagrammos lagocephalus*
Masked greenling *Hexagrammos octogrammus*

SCULPINS 49
Grunt sculpin *Rhamphocottus richardsonii*
Roughspine sculpin *Triglops macellus*
Scissortail sculpin *Triglops forficatus*
Red Irish lord *Hemilepidotus hemilepidotus*
Yellow Irish lord *Hemilepidotus jordani*
Threaded sculpin *Gymnocanthus pistilliger*
Armorhead sculpin *Gymnocanthus galeatus*
Leister sculpin *Enophrys lucasi*
Brightbelly sculpin *Microcottus sellaris*
Shorthorn sculpin *Myoxocephalus scorpius*
Plain sculpin *Myoxocephalus jaok*
Great sculpin *Myoxocephalus polyacanthocephalus*
Bigmouth sculpin *Hemitripterus bolini*
Crested sculpin *Blepsias bilobus*
Silverspotted sculpin *Blepsias cirrhosus*
Sailfin sculpin *Nautichthys oculofasciatus*
Spinyhead sculpin *Dasycottus setiger*
Darkfin sculpin *Malacocottus zonurus*

POACHERS 63
Sawback poacher *Leptagonus frenatus*
Fourhorn poacher *Hypsagonus quadricornis*
Sturgeon poacher *Podothecus accipenserinus*
Blackfin poacher *Bathyagonus nigripinnis*
Alligatorfish *Aspidophoroides monopterygius*

LUMPSUCKER 66
Smooth lumpsucker *Aptocyclus ventricosus*

SNAILFISHES 68
Variegated snailfish *Liparis gibbus*
Salmon snailfish *Careproctus rastrinus*
Spectral snailfish *Careproctus phasma*
Blotched snailfish *Crystallichthys cyclospilus*
Combed snailfish *Allocareproctus kallaion*
Cherry snailfish *Allocareproctus jordani*

ARMORHEAD 73
North Pacific armorhead *Pseudopentaceros wheeleri*

RONQUILS 74
Searcher *Bathymaster signatus*
Alaskan ronquil *Bathymaster caeruleofasciatus*

EELPOUTS 76
Shortfin eelpout *Lycodes brevipes*
Wattled eelpout *Lycodes palearis*
Polar eelpout *Lycodes turneri*
Alaska eelpout *Bothrocara pusillum*

PRICKLEBACKS 80
Decorated warbonnet *Chirolophis decoratus*
Longsnout prickleback *Lumpenella longirostris*
Snake prickleback *Lumpenus sagitta*
Whitebarred prickleback *Poroclinus rothrocki*
Blackline prickleback *Acantholumpenus mackayi*
Stout eelblenny *Anisarchus medius*

WRYMOUTHS 84
Giant wrymouth *Cryptacanthodes giganteus*
Dwarf wrymouth *Cryptacanthodes aleutensis*

GUNNEL 86
Crescent gunnel *Pholis laeta*

WOLFFISH 86
Bering wolffish *Anarhichas orientalis*

PROWFISH 88
Prowfish *Zaprora silenus*

SANDFISH 88
Pacific sandfish *Trichodon trichodon*

SAND LANCE 90
Pacific sand lance *Ammodytes hexapterus*

RIGHTEYE FLOUNDERS 90
Pacific halibut *Hippoglossus stenolepis*
Greenland turbot *Reinhardtius hippoglossoides*
Arrowtooth flounder *Atheresthes stomias*
Kamchatka flounder *Atheresthes evermanni*
Flathead sole *Hippoglossoides elassodon*
Butter sole *Isopsetta isolepis*
Rex sole *Glyptocephalus zachirus*
Southern rock sole *Lepidopsetta bilineata*
Northern rock sole *Lepidopsetta polyxystra*
Dover sole *Microstomus pacificus*
Yellowfin sole *Limanda aspera*
English sole *Parophrys vetulus*
Starry flounder *Platichthys stellatus*
Alaska plaice *Pleuronectes quadrituberculatus*

SPONGES **103**
GLASS SPONGE **103**
Vase sponge *Aulosaccus schulzei*
CALCAREOUS SPONGE **104**
Hairy vase sponge *Leucandra* sp.
VARIABLE SKELETON SPONGES **104**
Soccer ball sponge *Geodia mesotriaena*
Fuzzy cratered sponge *Vulcanella* sp.
Furry ball sponge *Craniella spinosa*
Baseball sponge *Craniella* sp.
Hermit sponges *Suberites ficus*, *Suberites domuncula*
Nipple sponge *Polymastia* sp.
Tree sponge *Mycale loveni*
Cotton ball sponge *Mycale lobata*
Lattice sponge *Melonchela clathriata*
Spud sponge *Histodermella* sp.
Firm fingered sponge *Neoesperiopsis* sp.
Rough China hat sponge *Neoesperiopsis infundibula*
Soft brown sponge *Myxilla* sp.
Scallop sponge *Myxilla incrustans*
HYDROIDS **114**
Champagne flute hydroid *Bonneviella* sp.
Ostrich-plume hydroid *Aglaophenia* sp.
HYDROCORALS **116**
Stylaster hydrocoral *Stylaster campylecus parageus*
Common hydrocoral *Stylaster campylecus campylecus*
Pink hydrocoral *Stylaster verrillii*
Capped hydrocoral *Crypthelia trophostega*
Rosy-undulate hydrocoral *Cyclohelia lamellata*
Rough hydrocoral *Errinopora zarhyncha*
OCTOCORALS (SOFT CORALS) **120**
Red mushroom coral *Anthomastus* sp.
Gray mushroom coral *Anthomastus* sp.
Sea raspberry *Gersemia* sp.
Purple soft coral *Anthothela grandiflora*
Bubblegum coral *Paragorgia arborea*
Sea fan *Calcigorgia* sp.
Purple sea fan *Muriceides nigra*
Kool's candelabrum gorgonian *Cryogorgia koolsae*
Pacific red sea fan *Swiftia* sp.

Octocorals (continued)
Pinnate amphilaphis *Amphilaphis* sp.
Plumarella coral *Plumarella* sp.
Pacific red tree coral *Primnoa pacifica*
Utinomi's armored sea fan *Arthrogorgia utinomii*
Superb thouarella *Thouarella superba*
Giant fanellia *Fanellia compressa*

CUP CORAL **130**
Alaska cup coral *Caryophyllia alaskensis*

SEA PEN **130**
Orange sea pen *Ptilosarcus gurneyi*

SEA WHIP **130**
Giant sea whip *Halipteris willemoesi*

SEA ANEMONES **132**
Mottled anemone *Urticina crassicornis*
Chevron-tentacled anemone *Cribrinopsis fernaldi*
Gigantic (white-plumed) anemone *Metridium farcimen*

JELLYFISH **134**
Sunrise jellyfish *Chrysaora melanaster*

SEGMENTED WORMS **136**
Sea mouse *Aphrodita negligens*
Depressed scale worm *Eunoe depressa*
Giant scale worm *Eunoe nodosa*
Red-banded commensal worm *Arctonoe vittata*
Northern feather duster worm *Eudistylia vancouveri*
Northern iridescent worm *Hemipodus borealis*
Calcareous tube worm *Serpula vermicularis*

LEECH **140**
Striped sea leech *Notostomum cyclostomum*

CHITONS **140**
Black Katy *Katharina tunicata*
Giant Pacific (gumboot) chiton *Cryptochiton stelleri*
White veiled chiton *Placiphorella pacifica*
Bering chiton *Lepidozona beringiana*

LIMPETS **144**
Rough keyhole limpet *Diodora aspera*
Whitecap limpet *Acmaea mitra*

SLIPPERSNAIL **144**
Great slippersnail *Crepidula grandis*

External Features of Some Gastropods **146**

MOONSNAILS **147**

Pale moonsnail *Euspira pallida*
Rusty moonsnail *Cryptonatica russa*
Aleutian moonsnail *Cryptonatica aleutica*
Arctic moonsnail *Cryptonatica affinis*

SNAILS **148**

Two-keel hairysnail *Trichotropis bicarinata*
Oregon triton *Fusitriton oregonensis*
Frilled dogwinkle *Nucella lamellosa*
Crown trophon *Boreotrophon coronatus*
Sinuous whelk *Buccinum plectrum*
Ladder (silky) whelk *Buccinum scalariforme*
Polar whelk *Buccinum polaris*
Angular whelk *Buccinum angulosum*
Helmet whelk *Clinopegma magnum*
Thick-cord whelk *Beringius crebricostatus*
Kennicott's beringius *Beringius kennicottii*
"Golden" beringius *Beringius* sp.
Bering's beringius *Beringius behringi*
Northern neptune *Neptunea heros*
Lyre neptune *Neptunea lyrata*
Fat whelk *Neptunea ventricosa*
Pribilof whelk *Neptunea pribiloffensis*
Little neptune *Neptunea communis borealis*
"Vesteraalen" neptune *Neptunea* sp.
Large melon whelk *Pyrulofusus melonis*
Left-hand melon whelk *Pyrulofusus harpa*
Warped whelk *Pyrulofusus deformis*
Tulip whelk *Volutopsius middendorffii*
Fragile whelk *Volutopsius fragilis*
Frilled whelk *Volutopsius trophonius*
Volute whelk *Volutopsius castanea*
Threaded whelk *Lussivolutopsius filosus*
Arctic whelk *Colus kroyeri*
Oblique whelk *Colus hypolispus*
Shrew whelk *Colus hallii*
Ribbed volute *Arctomelon tamikoae*
Alaska volute *Arctomelon stearnsii*

SNAIL EGGS **170**
Fat neptune eggs *Neptunea ventricosa* eggs
Lyre neptune eggs *Neptunea lyrata* eggs
Northern neptune eggs *Neptunea heros* eggs
Buccinum eggs *Buccinum* sp. eggs
Beringius eggs *Beringius* sp. eggs
Oregon triton eggs *Fusitriton oregonensis* eggs

NUDIBRANCHS **172**
Pacific sea-lemon *Anisodoris nobilis*
"Spotted" diaulula *Diaulula* sp.
Giant orange tochui *Tochuina tetraquetra*
Tritonid nudibranch *Tritonia* sp.
Sea-clown triopha *Triopha catalinae*
Dall's dendronotid *Dendronotus dalli*
Red frond-aeolis *Dendronotus rufus*
Lion nudibranch *Melibe leonina*
Golden dirona *Dirona aurantia*
Predaceous aeolis *Flabellina trophina*
Shag-rug aeolis *Aeolidia papillosa*

MUSSELS **178**
Northern horsemussel *Modiolus modiolus*
Black mussel *Musculus niger*
Discordant mussel *Musculus discors*
Foolish mussel *Mytilus trossulus*

OYSTER **182**
Pacific oyster *Crassostrea gigas*

JINGLE **182**
Alaska (green false) jingle *Pododesmus macroschisma*

SCALLOPS **182**
Reddish scallop *Chlamys rubida*
Weathervane scallop *Patinopectin caurinus*

External Features of Cockles and Clams **185**

COCKLES **186**
Hairy cockle *Clinocardium ciliatum*
Nuttall cockle *Clinocardium nuttallii*
Greenland smoothcockle *Serripes groenlandicus*
Oblique smoothcockle *Serripes notabilis*
Broad smoothcockle (broad cockle) *Serripes laperousii*

CLAMS **188**
Washington butterclam *Saxidomus gigantea*
Pacific littleneck *Protothaca staminea*
Yoldia *Yoldia* sp.
Broad yoldia *Megayoldia thraciaeformis*
Disjunct cleft clam *Conchocele disjuncta*
Giant cleft clam *Conchocele bisecta*
Elegant soft shell *Mya elegans*
Alaska great-tellin *Tellina lutea*
Arctic roughmya *Panomya norvegica*
Arctic hiatella *Hiatella arctica*
Pilsbry piddock *Zirfaea pilsbryi*
Alaska razor clam *Siliqua alta*
Pacific razor clam *Siliqua patula*
Arctic surfclam *Mactromeris polynyma*
Fat gaper clam *Tresus capax*
Arctic astarte *Astarte arctica*
Elliptical astarte *Astarte elliptica*

BRACHIOPODS **200**
Black lamp shell *Hemithiris psittacea*
Transverse lamp shell *Terebratalia transversa*

SQUIDS **202**
Magister armhook squid *Berryteuthis magister*
North Pacific bobtail squid *Rossia pacifica*

OCTOPUS **202**
Pacific giant octopus *Enteroctopus dofleini*

SEA SPIDER **204**
Giant sea spider *Colossendeis* sp.

BARNACLES **204**
Eared barnacle *Tarasovium cornutum*
Pelagic gooseneck barnacle *Lepas* sp.
Beaked barnacle *Balanus rostratus*
Giant barnacle *Chirona evermanni*
Crab barnacle *Hesperibalanus hesperius*

ISOPODS **208**
Gravel isopod *Synidotea* sp.
Sea cockroach *Rocinela angustata*
Spiky arcturid *Arcturus* sp.
Spiny arcturid *Arcturus* sp.

AMPHIPOD **210**
Skeleton shrimp *Caprella* sp.

SHRIMPS **211**
Shrimp Anatomy **211**
Sidestripe shrimp *Pandalopsis dispar*
Aleutian bigeye *Pandalopsis aleutica*
Spot shrimp (prawn) *Pandalus platyceros*
Coonstripe shrimp *Pandalus hypsinotus*
Dock shrimp *Pandalus danae*
Alaskan pink shrimp *Pandalus eous*
Humpy shrimp *Pandalus goniurus*
Yellowleg pandalid *Pandalus tridens*
Spiny lebbeid *Lebbeus groenlandicus*
Arctic argid *Argis dentata*
Sculptured shrimp *Sclerocrangon boreas*

CRABS **220**
External Features of Crabs **220**
Puget sound king crab *Lopholithodes mandtii*
Brown box crab *Lopholithodes foraminatus*
Rhinoceros crab *Rhinolithodes wosnessenskii*
Heart crab *Phyllolithodes papillosus*
Spiny lithode crab *Acantholithodes hispidus*
Soft crab *Hapalogaster grebnitzkii*
Hairy crab *Hapalogaster mertensii*
Butterfly crab *Cryptolithodes typicus*
Red king crab *Paralithodes camtschaticus*
Blue king crab *Paralithodes platypus*
Spiny king crab *Paralithodes brevipes*
Golden king crab *Lithodes aequispinus*
Scarlet king crab *Lithodes couesi*
Scaled crab *Placetron wosnessenskii*
Verrill's spider crab *Paralomis verrilli*
Spiny paralomis *Paralomis multispina*
Spiny spider crab *Paralomis makarovi*
Pinchbug *Munida quadrispina*
Triangle Tanner crab *Chionoecetes angulatus*
Grooved Tanner crab *Chionoecetes tanneri*
Tanner crab *Chionoecetes bairdi*
Snow crab *Chionoecetes opilio*
Hybrid Tanner crab *Chionoecetes bairdi × opilio*
Graceful decorator crab *Oregonia gracilis*

Pacific lyre crab *Hyas lyratus*
Arctic lyre crab *Hyas coarctatus*
Graceful kelp crab *Pugettia gracilis*
Longhorn decorator crab *Chorilia longipes*
Hair crab *Erimacrus isenbeckii*
Helmet crab *Telmessus cheiragonus*
Dungeness crab *Cancer magister*
Pygmy rock crab *Cancer oregonensis*
Gaper pea crab *Pinnixa littoralis*

HERMIT CRABS **242**
Hermit Crab Anatomy **242**
Splendid hermit *Labidochirus splendescens*
Widehand hermit *Elassochirus tenuimanus*
Purple hermit *Elassochirus cavimanus*
Pacific red hermit *Elassochirus gilli*
Aleutian hermit *Pagurus aleuticus*
Alaskan hermit *Pagurus ochotensis*
Longfinger hermit *Pagurus rathbuni*
"Dirty harry" hermit *Pagurus capillatus*
Fuzzy hermit *Pagurus trigonocheirus*
Whiteknee hermit *Pagurus dalli*
Knobbyhand hermit *Pagurus confragosus*

BRYOZOANS **250**
Jointed bryozoan *Microporina borealis*
Fan bryozoan *Dendrobeania* sp.
Spiral bryozoan *Bugula californica*
Smooth leather bryozoan *Alcyonidium pedunculatum*
Medusa bryozoan *Alcyonidium* sp.
Spiny leather bryozoan *Flustrellidra corniculata*

FEATHER STAR **254**
Feather star *Florometra* sp.

SEA STARS **255**
External Features of Sea Stars **255**
Arctic sand star *Leptychaster arcticus*
Common mud star *Ctenodiscus crispatus*
Fragile star *Cheiraster dawsoni*
Orange cookie star *Ceramaster patagonicus*
Arctic cookie star *Ceramaster arcticus*
Stellate cookie star *Ceramaster stellatus*
Red cookie star *Ceramaster japonicus*

Sea stars (continued)

Grainy white cookie star *Cladaster validus*
Kurile spiny star *Hippasteria kurilensis*
Pale spiny star *Hippasteria* sp.
Northern scarlet star *Pseudarchaster parelii*
Alaskan scarlet star *Pseudarchaster parelii alascensis*
Leather star *Dermasterias imbricata*
Striped sun star *Solaster stimpsoni*
Remarkable sun star *Solaster* sp.
Beautiful sun star *Solaster* sp.
Serpent sun star *Solaster* sp.
Rose star *Crossaster papposus*
Crested star *Lophaster furcilliger*
Tessellated slime star *Pteraster tesselatus*
Northern slime star *Pteraster* sp.
Obscure cushion star *Pteraster obscurus*
Eight-rayed cushion star *Pteraster octaster*
Wrinkled cushion star *Pteraster militaris*
Pincushion star *Diplopteraster multipes*
Blood star *Henricia* cf. *leviuscula*
Spiny henricia *Henricia multispina*
Tumid henricia *Henricia tumida*
Bering henricia *Henricia beringiania*
Fuzzy henricia *Henricia* sp.
Purple toothed henricia *Odontohenricia* sp.
Pale toothed henricia *Odontohenricia* sp.
Purple-orange star *Asterias amurensis*
Giant star *Evasterias echinosoma*
Mottled star *Evasterias troschelii*
Arctic star *Leptasterias arctica*
Aleutian star *Leptasterias hylodes*
Knobby six-rayed star *Leptasterias polaris*
Giant Aleutian six-rayed star *Leptasterias truculenta*
Sunflower star *Pycnopodia helianthoides*
Redbanded star *Orthasterias koehleri*
Blackspined star *Lethasterias nanimensis*
Long-rayed star *Stylasterias forreri*
Ochre star *Pisaster ochraceus*

BASKET STAR **284**
Basket star *Gorgonocephalus eucnemis*

SERPENT AND BRITTLE STARS **286**
External Features of Serpent and Brittle Stars **286**
Dwarf smooth serpent star *Astronebris tatafilius*
Giant serpent star *Asteronyx loveni*
White serpent star *Astrochele laevis*
Daisy brittle star *Ophiopholis aculeata kennerlyi*
Ubiquitous brittle star *Ophiopholis aculeata aculeata*
Spiny-brown brittle star *Ophiacantha cataleimmoidus*
Nine-armed brittle star *Ophiacantha enneactis*
Nodulose brittle star *Ophiolebes* sp.
Notched brittle star *Ophiura sarsi*
Great armored brittle star *Stegophiura ponderosa*

SEA URCHINS **294**
Aleutian urchin *Strongylocentrotus polyacanthus*
Green urchin *Strongylocentrotus droebachiensis*
Red urchin *Strongylocentrotus franciscanus*
Fragile (pink) urchin *Strongylocentrotus fragilis*
Heart urchin *Brisaster latifrons*

SAND DOLLAR **296**
Northern (Parma) sand dollar *Echinarachnius parma*

SEA CUCUMBERS **298**
Giant California cucumber *Parastichopus californicus*
Northern papillate (slender) cucumber *Bathyplotes* sp.
Red-orange armored cucumber *Psolus japonicus*
White-scaled cucumber *Psolus squamatus*
Arctic armored cucumber *Psolus fabricii*
Sea football *Cucumaria fallax*
Crescent cucumber *Pentamera lissoplaca*
Sweet sea potato *Molpadia intermedia*

TUNICATES **302**
Sea onion *Boltenia ovifera*
Bristly tunicate *Halocynthia hilgendorfi igaboja*
Sea peach *Halocynthia aurantium*
Sea potato *Styela rustica*

LITERATURE 306
PHOTO TECHNIQUES AND CREDITS 310
GLOSSARY 312
PHYLOGENIC SPECIES LIST 320
INDEX TO COMMON AND SCIENTIFIC NAMES........ 327
PICTURE INDEX 336

ACKNOWLEDGMENTS

The authors gratefully thank the following individuals for their help in making this book a reality. As long-time at-sea biologists, we are particularly indebted to Doyne Kessler for his handy and thorough *Alaska's Saltwater Fishes and Other Sea Life: A Field Guide* published in 1985 which, especially during the early part of our careers, afforded us the ability to quickly learn and identify those species common to survey catches. Production of this guide would not have been possible without the thorough reviews, species confirmations, and suggested additions given by James Orr, Bruce Wing, R.A. MacIntosh, and Bob Stone (National Marine Fisheries Service), Greg Morgan (Observer Training Center, University of Alaska Anchorage), and Elaina Jorgensen (NMFS Alaska Fisheries Science Center). Ryan Burt, Kevin Renfro, and Melissa Salmon (Alaska Department of Fish and Game) assisted in the identification and photography of the marine organisms in this guide. We greatly appreciate the comments given by four anonymous reviewers; critical changes in guide content were made as a result of their contributions. We are especially indebted to Roger Clark for producing the draft copies of his invertebrate guide that we used as a primary source for many of the animals in the guide. We also thank Roger for graciously making critical edits and providing new information and photographs to us for inclusion in this guide. Special thanks go to Wayne Donaldson, Ryan Burt, and Lisa Marcato (ADFG) for editing, and Janelle Peterson, for layout and production. And last, Douglas Pengilly (ADFG) is gratefully acknowledged for securing long-term funding for our positions.

PROJECT SPONSORSHIP

Preparation of this guide was funded in part by a cooperative agreement from the National Oceanic and Atmospheric Administration (NOAA) under NOAA Awards NA05NMF4371036, NA06NMF4370073, and NOA7NMF4370268. Funding for layout and production was provided by the Pacific States Marine Fisheries Commission through the Alaska Department of Fish and Game. The views expressed herein are those of the authors and do not necessarily reflect the views of NOAA or any of its subagencies.

ABOUT THE AUTHORS

Susan C. Byersdorfer has been a shellfish research biologist for the Alaska Department of Fish and Game in Kodiak, Alaska, since 1989. Susan's principal work has been on the life history of several commercially important marine species including Dungeness, king, and Tanner crabs, as well as black rockfish. Her primary work has involved crab population surveys and long-term tagging studies of red, blue, and golden king crabs in the Bering Sea and Aleutian Islands. She has also focused on black and dark rockfish tagging and fecundity studies, ageing of horse clams, and bitter crab research. She is the author and coauthor of more than 35 published papers and journal articles, and recently coauthored the book *Biological Field Techniques for Lithodid Crabs*.

Leslie J. Watson was a marine research biologist for 26 years with the Alaska Department of Fish and Game in Kodiak. She has worked on commercially important salmon, groundfish, shrimp, crabs, and other invertebrates since the mid-1970s by conducting vessel surveys, gear, and laboratory studies. Most recently she has carried out long-term, large-scale red, blue, and golden king crab tagging projects in the Bering Sea and Aleutian Islands. She is the author or coauthor of more than 75 published papers and journal articles.

INTRODUCTION

Annual trawl and pot surveys have been conducted by the Alaska Department of Fish and Game and National Marine Fisheries Service in the Gulf of Alaska, Bering Sea, and Aleutian Islands waters for decades. Although the primary purpose of such surveys has been to determine stock size and condition for commercially important fish and crab species, the biological composition of trawl and pot catches is an integral part of each survey and has provided valuable information on community structure and species ranges.

While several excellent guides and keys have been useful for species identification in the field, the need for an updated simple, annotated photographic guide covering most of the common marine biota encountered during surveys is both desirable and warranted. This guide, while not all inclusive, attempts to fill that need and is of primary use to at-sea biologists and technicians, but also to fishers, teachers, and anyone who just wants to know what they've caught. The primary intent of this guide is to provide a clear, concise, first stop for the correct identification of marine life. The user is cautioned that many animals are superficially similar in appearance, and that other identification resources may be necessary to distinguish similar animals. Certainly, this guide does not contain all of the common animals found in all surveys conducted in Alaska waters. For that reason we have provided users of this guide a broad selection of available references for attaining positive identifications of captured species.

Marine life selected for inclusion in the guide represents animals most commonly found in our ADFG pot and trawl surveys. In some cases, especially involving invertebrates, only the genus is reported because the species has yet to be confirmed. Other gaps in representation are simply the result of non-captured or rarely captured specimens, or more often, unconfirmed identifications of animals. For those gaps, other common species are noted within families or genera with descriptive characteristics that may aid in the identification of animals not found in the guide.

The primary geographic range encompassed by the guide includes the waters of the eastern Bering Sea, central Aleutian Islands, western and central Gulf of Alaska, and to a lesser degree, those areas north of Norton Sound, Alaska, and southeastern Alaska. Where applicable, reference to broader distributions beyond Alaska waters is given, i.e., circumboreal. Given range descriptions may also include extensions south to the continental coast of North America, and west to the Kamchatka Peninsula of Russia.

Taxonomic classification of fishes and invertebrates is given for reference purposes only. In many cases only higher taxa are given. The classification of fishes is based on Mecklenburg et al. (2002) and the classification of invertebrates is loosely patterned after O'Clair and O'Clair (1998), and to a lesser extent, Kozloff (1987).

All classifications were verified using the Integrated Taxonomic Information System (2008). We relied heavily on R. Clark's 2006 unpublished manuscript, "Field Guide to the Benthic Marine Invertebrates of Alaska's Shelf and Upper Slope," for identification of miscellaneous invertebrates and as such, take full responsibility for any errors within this guide that arise from its use.

In the text, several groupings are not in phylogenic order—most notably the lamprey, shark, skate, and fish sections. These were placed at the beginning of the guide as an aid to users, since these animals are commonly noted first. Brachiopods (lamp shells), though not closely related to the bivalves, are nonetheless placed immediately following the clams since both groups have paired calcareous shells. Salmon and trout precede herring and smelts as in Kessler (1985). Lastly, the bryozoan grouping simply follows that of the hermit crabs based on the order found in Kozloff (1987). Refer to the Phylogenic Species List for a suggested phylogeny of the animals contained in the guide; however, readers should be aware that there are other equally valid systems in use by taxonomists.

Our reporting of scientific names within the guide reflects nomenclature gathered from many sources. For sharks and skates we relied on Stevenson et al. (2007) and Mecklenburg et al. (2002), for fishes Mecklenburg et al. (2002) and Kessler (1985), and for invertebrates we primarily employed Clark (2006), Kessler (1985), Lambert and Austin (2007), and Lambert (1997 and 2000). When a scientific name was recently changed or not agreed upon, the alternate name is noted in the description, e.g., for arrowtooth flounder *Atheresthes stomias*, the name *Reinhardtius stomias* is also accepted.

Verifying the best common names, especially for invertebrates, was problematic due to the variety of common names used by different authors. Where longstanding common names were not apparent, we first polled the American Fisheries Society publications by Turgeon et al. (1998) and McLaughlin et al. (2005). When a verified common name did not exist, we used the referenced author's common name. Further, when the referenced author used a provisional common name, we reported that using quotation marks, e.g., "golden" beringius. When more than one common name is frequently used for a single animal, we have enclosed one of the two names in parentheses, e.g., sockeye (red) salmon.

HOW TO USE THIS GUIDE

Identification of animals in the guide can be accomplished using a combination of characteristics such as color or form, and by inclusion of only that information necessary to positively identify an animal. In some cases, animals may only be distinguished by slight differences; for that reason, a combination of written descriptions, photographs, and diagrams must be used to ensure positive identifications.

Using the step-wise instructions found in Kessler (1985), first consult the Picture Index on page 336 to determine the grouping that your animal most closely resembles. Next, turn to the corresponding page number for that diagram and scan through the photographs to find the one that best fits the most apparent features of your animal. At this point, you may have narrowed your search to two or more possible species. By using the descriptive text, you may now determine which species description applies to your animal. Make sure that the description and the accompanying photograph provide the only possible positive identification of your specimen. If there is not a match between the animal in question and any of the available text descriptions and their photos, reexamine the animal and run through the identification process again. If the animal is not found upon reexamination, refer to one of the appropriate keys or guides listed in the Literature section for more identification options. When all available avenues of species identification have been pursued and the animal has not been identified, or if you are unsure of the resulting identification, freeze the animal and enclose a label stating the date, location, depth, and other information related to its capture. Return the frozen specimen to the Alaska Department of Fish and Game or National Marine Fisheries Service office nearest to your location for future identification.

LAMPREY

Arctic lamprey

Lethenteron camtschaticum

DESCRIPTION: Blue-black to dark brown dorsally, yellow to light brown ventrally; dorsal fins light tan to gray; dark blotch on second dorsal fin and on tail. Supraoral bar with 2 cusps, rarely 3 (1). Infraoral bar with 5-10 cusps, usually 7 or 8 (2). Lateral tooth plates 3, central pair with 2 cusps (3). Posterior teeth present. The Pacific lamprey *Lampetra tridentata* is the only other lamprey found in Alaska waters, and differs from the arctic lamprey by having 3 cusps on the supraoral bar, 4 pairs of lateral tooth plates, and a smaller geographic range.
SIZE: Length to 63 cm TL.
DISTRIBUTION: Arctic coast to Kenai Peninsula; up Yukon River into Yukon Territory, also in Kuskokwim and Tanana river drainages; almost circumpolar from Lapland south to Caspian Sea, eastward to Kamchatka and south to Korea. Anadromous, parasitic; at sea to depth of 50 m.
(Mecklenburg et al. 2002)

SHARKS

Pacific sleeper shark

Somniosus pacificus

DESCRIPTION: Uniform blackish brown to slate green or gray, sometimes slightly streaked. A large, sluggish shark, it is distinguished from the spiny dogfish by the lack of dorsal fin spines. Dorsal fins low, nearly equal in size. Anal fin absent; caudal fin somewhat paddle-shaped, with large lower lobe. Body robust and flaccid; snout long and rounded. Upper teeth slender-cusped, narrow, thornlike; cusps of lower teeth short and low, strongly oblique, and roots very high.
SIZE: Documented length to 4.4 m TL.
DISTRIBUTION: Bering Sea, Aleutian Islands, and Gulf of Alaska. Benthic, intertidal to 2,000 m depth.
(Kessler 1985, Mecklenburg et al. 2002, Stevenson et al. 2007)

Lethenteron camtschaticum

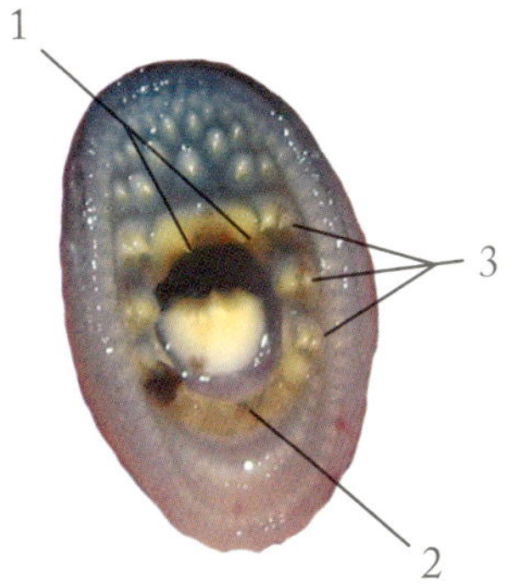

Somniosus pacificus

Spiny dogfish

Squalus acanthias

DESCRIPTION: Slate gray or brown dorsally, lighter ventrally. Dorsal precaudal pit present; lateral keel present on caudal peduncle. First dorsal fin larger than second; single spine at origin of each dorsal fin (1); anal fin absent; 5 gill openings (2).
SIZE: Length to 2 m TL.
DISTRIBUTION: Bering Sea and Gulf of Alaska, with one record from Kotzebue Sound. Usually benthic, intertidal to 1,244 m depth.
(Mecklenburg et al. 2002, Stevenson et al. 2007)

Squalus acanthias

OTHER SHARKS. The other common large shark found in Alaska waters is the salmon shark *Lamna ditropis*, a high-value sport fish that is occasionally caught in nets and trawls.

SKATES

EXTERNAL FEATURES OF SKATES

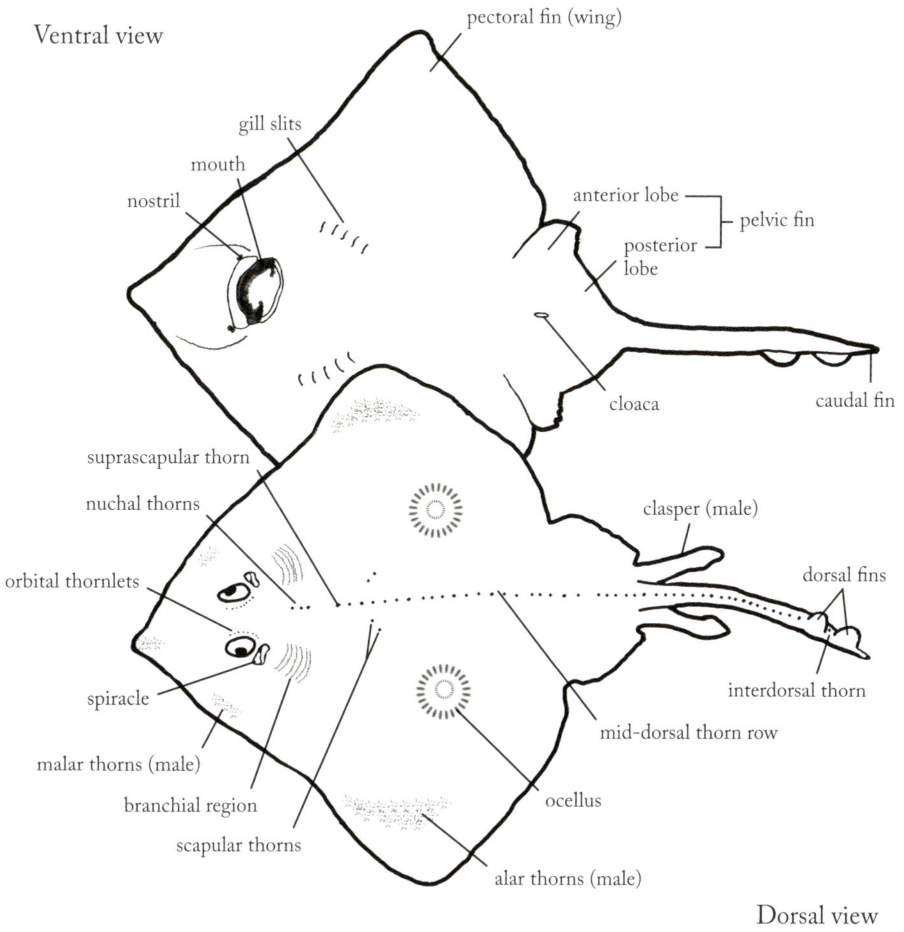

Skate egg case

Genus *Raja*: snout stiff, separated from wing support cartilage by a wide space (1). Wing margin concave, i.e., a straight line from wingtip to snout is mostly outside margin of wing (2).

Longnose skate

Raja rhina

DESCRIPTION: Dorsal surface is brown with dark blotches, may have ocelli on disk, and is covered with fine denticles; ventral surface smooth and bluish gray or mottled brown to black in color. Deeply notched pelvic fins. Snout stiff and elongate. Middorsal row of thorns on tail only, except for 1 or 2 nuchal thorns; scapular thorns absent; orbital thorns present.
SIZE: Length to about 142 cm TL.
DISTRIBUTION: Eastern Bering Sea and Gulf of Alaska to Baja California. Depth 0-169 m.
(Kessler 1985, Mecklenburg et al 2002, Stevenson et al. 2007)

Big skate

Raja binoculata

DESCRIPTION: Dorsal surface brown, gray-brown, or reddish brown, often with large, dark-centered ocellus surrounded by smaller spots on pectoral fins. Ventral surface is white to light gray, sometimes with light blotches. Snout stiff, long, bluntly pointed with shallowly notched pelvic fins. One middorsal thorn in nuchal area, continuous middorsal row starts near pelvic region.
SIZE: Length to 244 cm TL.
DISTRIBUTION: Eastern Bering Sea, eastern Aleutian Islands, Gulf of Alaska to Baja California. Depth 16-402 m.
(Kessler 1985, Mecklenburg et al. 2002, Stevenson et al. 2007)

Raja rhina

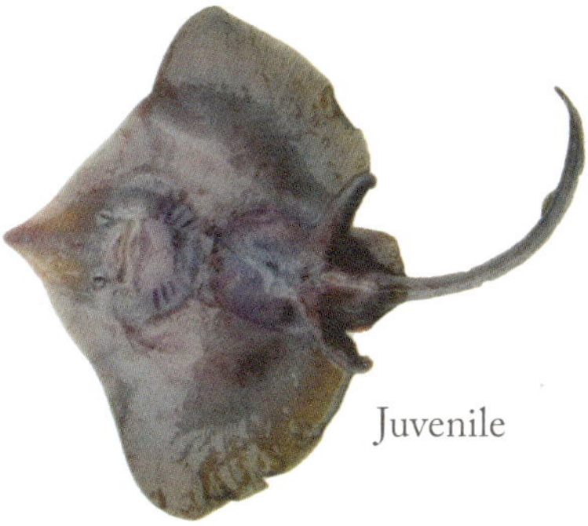
Juvenile

Raja binoculata

Genus *Bathyraja*: snout flexible, separated from wing support cartilage by a narrow gap. Wing margin undulate, i.e., a straight line from wingtip to snout is mostly inside margin of wing (1).

Alaska skate

Bathyraja parmifera

DESCRIPTION: Dark grayish brown to olive brown dorsally with black spots, sometimes with variable, paired light spots; white ventrally, with dark blotches on tail. Snout flexible, blunt; disk width 65% of total length or greater. Wing margin undulate (1). Scapular thorns strong, usually 2 on each side; orbital thornlets present (2). The only species with naked area around tail thorns. Row of middorsal thorns may be complete or interrupted. Tail length shorter than precaudal length.
SIZE: Length to 135 cm TL.
DISTRIBUTION: Bering Sea, Aleutian Islands, Gulf of Alaska; western Bering Sea to Japan. Depth 17-392 m.
(Mecklenburg et al. 2002, Stevenson et al. 2007)

"Leopard" skate

Bathyraja sp. cf. *parmifera*

DESCRIPTION: Dorsal surface with vivid olive-yellow leopard spotting pattern; ventral surface gray to white. The "leopard" skate is currently an undescribed species that appears to be restricted in Alaska waters to the Aleutian Islands. Recent genetic studies indicate it is a distinct species closely related to the Alaska skate.
SIZE: Length to 139 cm TL.
DISTRIBUTION: Western Aleutian Islands. Depth 48-396 m.
(Stevenson et al. 2007)

Mud skate

Bathyraja taranetzi

DESCRIPTION: Dorsal surface brown with small darker blotches and usually a large, lighter spot on posterior part of pectoral fin; ventral surface white or cream colored. In the Aleutian Islands, dorsal surface with distinctive yellow spots and blotches. Tail length longer than precaudal length in adults. No scapular thorns; middorsal nuchal thorns usually absent. Tail thorns unevenly distributed, usually decrease to denticles dorsally. Dorsal surface covered with denticles; ventral surface smooth. Similar species: butterfly and Okhotsk skates. (Also known as *Rhinoraja taranetzi*.)
SIZE: Length to 77 cm TL.
DISTRIBUTION: Bering Sea and Aleutian Islands; western Bering Sea to northern Kuril Islands. Depth 58-1,054 m.
(Mecklenburg et al. 2002, Stevenson et al. 2007)

Bathyraja parmifera

Bathyraja sp. cf. *parmifera*

Bathyraja taranetzi

Aleutian skate

Bathyraja aleutica

DESCRIPTION: Dorsal surface grayish brown to dusky olive, with or without vague black ocellus on pectoral fin; ventral surface white with gray margin (1). Orbital thorns absent; scapular thorns strong; middorsal row continuous from nuchal area to first dorsal fin or interrupted over trunk; tail thorns large, with 1 or 2 strong thorns between dorsal fins (2). Fine denticles evenly distributed over dorsal surface; ventral surface smooth.

SIZE: Length to 161 cm TL.

DISTRIBUTION: Bering Sea, Aleutian Islands, Gulf of Alaska to northern California; western Bering Sea to Japan. Depth 15-1,602 m.

(Mecklenburg et al. 2002, Stevenson et al. 2007)

Bering skate

Bathyraja interrupta

DESCRIPTION: Dorsal surface brown to blackish, sometimes with white mottling; often with white spot on each side of tail; ventral surface white, underside of tail mostly dark. Row of middorsal thorns usually interrupted (1); tail thorns reduced, with 0-2 reduced thorns between dorsal fins; scapular thorns strong. Denticles densely cover dorsal surface, ventral surface smooth.

SIZE: Length to 86 cm TL.

DISTRIBUTION: Bering Sea and eastern Aleutian Islands, Gulf of Alaska to California. Depth 26-1,050 m.

(Mecklenburg et al. 2002, Stevenson et al. 2007)

Bathyraja aleutica

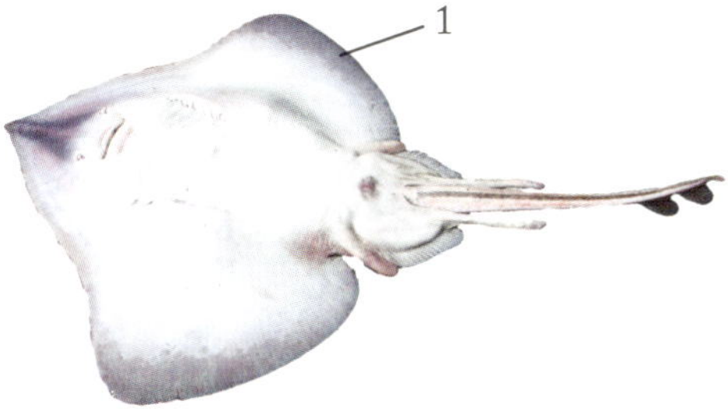

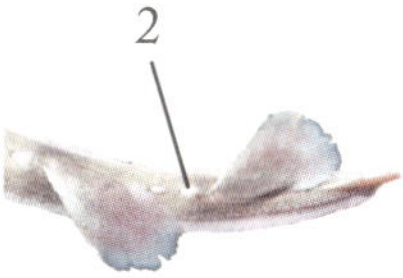

Bathyraja interrupta

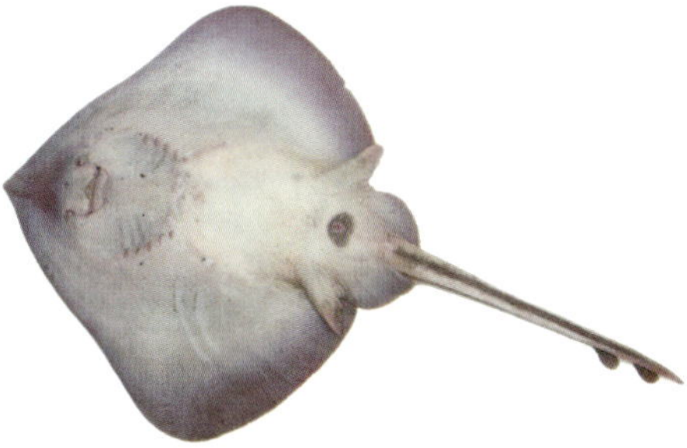

Commander skate

Bathyraja lindbergi

DESCRIPTION: Both surfaces dark grayish brown, ventral surface sometimes whitish brown; dorsal surface rarely with white mottling; whitish around mouth and distally on anterior lobe of pelvic fin (1). Scapular thorns absent; middorsal thorns more or less continuous from scapular region to first dorsal fin; thorns well developed along length of tail.
SIZE: Length to 102 cm TL.
DISTRIBUTION: Bering Sea, Aleutian Islands, and Gulf of Alaska; western Bering Sea to Japan. Depth 126-1,193 m.
(Mecklenburg et al. 2002, Stevenson et al. 2007)

Whiteblotched skate

Bathyraja maculata

DESCRIPTION: Dorsal surface gray, with large, scattered white to yellow blotches (1); ventral surface gray with darker tail and disk margins. Scapular thorns absent; nuchal thorns present, strong (2); no other middorsal thorns on disk; middorsal tail thorns strongly developed. Dorsal surface with strong denticles, ventral surface mostly smooth.
SIZE: Length to 147 cm TL.
DISTRIBUTION: Bering Sea and Aleutian Islands; western Bering Sea to Japan. Depth 73-1,193 m.
(Mecklenburg et al. 2002, Stevenson et al. 2007)

OTHER SKATES: There are 15 documented species of skates in Alaska. Only the nine that are commonly found in pot or trawl surveys are presented here. For other skates see the *Field Guide to Sharks, Skates, and Ratfish of Alaska* by Stevenson et al. 2007.

Bathyraja lindbergi

Bathyraja maculata

Female

EXTERNAL FEATURES OF SOFT-RAYED FISHES

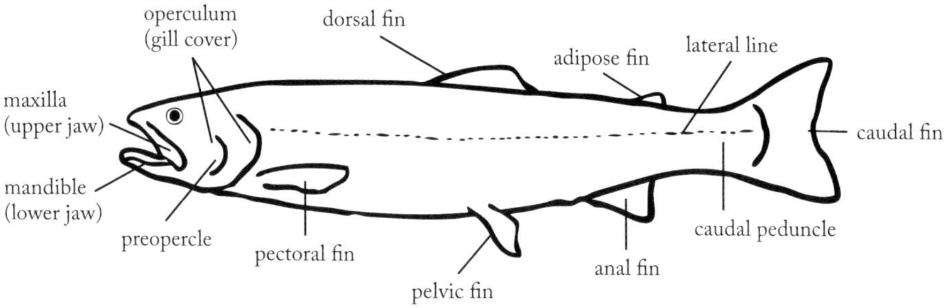

TROUTS

Rainbow trout

Oncorhynchus mykiss

DESCRIPTION: Dark blue to greenish or brown dorsally; silvery white or pale yellow ventrally; reddish stripe along side, except steelhead fresh from the sea; small dark spots on back, sides, and dorsal and caudal fins. Steelhead are the anadromous form of *O. mykiss* and resemble chinook salmon in saltwater coloration; white tongue and completely white mouth cavity in *O. mykiss* are distinguishing characteristics.
SIZE: Length to 122 cm TL.
DISTRIBUTION: Coastal waters and drainages from the Kuskokwim River and Port Moller to Mexico; introduced in interior Alaska and widely introduced elsewhere.
(Kessler 1985, Mecklenburg et al. 2002)

Cutthroat trout

Oncorhynchus clarkii

DESCRIPTION: Silvery to brassy with yellow tints; small, densely packed, irregularly shaped dark-brown or black spots on body, head, and fins; red or orange streak on underside of lower jaw on each side (1). Fewer spots and cutthroat marks absent or inconspicuous in fish at sea or fresh from sea; no red stripes on side. Maxilla extends well beyond eye; spawning males develop slight kype. Caudal fin slightly indented. Gill rakers short and blunt. This species is very difficult to separate from *O. mykiss* at the northwestern end of its range.
SIZE: Length to 99 cm TL.
DISTRIBUTION: Kenai Peninsula, Alaska, to California, including islands with suitable habitat in Gulf of Alaska. Anadromous and nonmigratory freshwater stocks. Widely stocked in lakes within and outside its original range.
(Mecklenburg et al. 2002)

Oncorhynchus mykiss

Steelhead

Oncorhynchus clarkii

Dolly Varden

Salvelinus malma

DESCRIPTION: Olive green to dark blue or brown dorsally; yellow, orange, red, or pink spots on side (1), none on tail; largest spots usually smaller than eye diameter (may be missing in saltwater phase); spots profuse. Leading edge of pelvic and anal fins cream-colored (2). Head slightly compressed, mandibular pores usually 6. Gill rakers strongly compressed, finely tapered. Number of branchiostegal rays 21-23.
SIZE: Length to 100 cm TL.
DISTRIBUTION: Arctic Alaska to northern Washington; anadromous, overwinter in lakes and spawn in tributaries to the lake, or resident in tributaries.
(Kessler 1985, Mecklenburg et al. 2002)

SALMON

All salmon have an adipose fin, a small finlike structure at the base of pelvic fin, dark tongues, and the rear edge of anal fin is slanted.

Pink (humpback) salmon

Oncorhynchus gorbuscha

DESCRIPTION: At sea, metallic blue to blue-green dorsally, silver laterally, white ventrally; large oval black spots on body dorsolaterally and on both lobes of caudal fin (1). Spawning males are dark on back, red with brownish green blotches on side; breeding females similar but less distinctly colored. Males develop pronounced hump and elongate, hooked snout and lower jaw (kype) during spawning. Scales are very small, usually more than 170 on lateral line; gums light.
SIZE: Length to 76 cm TL.
DISTRIBUTION: Beaufort Sea coast of Alaska to central California. During ocean phase, found throughout North Pacific and Bering Sea north of about 40°N. Anadromous; at sea, remain close to shore until 6-8 cm long, then move offshore.
(Mecklenburg et al. 2002)

Coho (silver) salmon

Oncorhynchus kisutch

DESCRIPTION: At sea, metallic blue dorsally, silver laterally, white ventrally; small black spots on body dorsolaterally and on upper lobe of caudal fin; gums at base of teeth in lower jaw white (1). Spawning male with dusky green back and head, bright red side, black belly; breeding female with bronze to pinkish red side. Scales large, 118-147 in first row above lateral line.
SIZE: Length to 108 cm TL.
DISTRIBUTION: Point Hope, Alaska, to central California. Anadromous; at sea move offshore and travel major current systems.
(Kessler 1985, Mecklenburg et al. 2002)

Salvelinus malma

Oncorhynchus gorbuscha

Oncorhynchus kisutch

Chinook (king) salmon

Oncorhynchus tshawytscha

DESCRIPTION: At sea, metallic dark green to blue-black dorsally, silver laterally, white ventrally; small black spots on both lobes of caudal fin (1) and on back; gums at base of teeth in lower jaw black (2). Spawning fish change to olive brown, red, or purplish color, with change more marked in males. Gill rakers rough, short, and widely spaced (3). Scales relatively large, 140-153 in first row above lateral line.
SIZE: Length to 160 cm TL.
DISTRIBUTION: Point Hope, Alaska, to California. Anadromous; at sea, some remain near shore but others travel out to sea and descend to 200 m or more.
(Mecklenburg et al. 2002)

Chum (dog) salmon

Oncorhynchus keta

DESCRIPTION: At sea, metallic dark blue dorsally, silver laterally and ventrally, no distinct black spots. Spawning fish become dirty red on sides, dusky below, and develop irregular purplish green bands on sides (1). Dorsal and anal fins and tail edged with black; slender tail base. Gill rakers smooth, fairly short, stout, widely spaced, fewer than 30.
SIZE: Length to 109 cm TL.
DISTRIBUTION: Beaufort Sea coast of Alaska to central California. Found close to shore, then disperse to open ocean and as deep as 61 m.
(Kessler 1985, Mecklenburg et al. 2002)

Sockeye (red) salmon

Oncorhynchus nerka

DESCRIPTION: At sea, metallic dark blue to greenish blue on head and back, silver laterally, white to silver ventrally. At spawning, male changes to bright olive green on head, brilliant red body, grayish to green paired fins and tail; females less brilliantly colored. Typically no distinct black spots on back or fins; if black spots present, they're small and limited to caudal or dorsal fin; tail base not slender (1). Gill rakers long, slender (2), closely spaced, 28-40 on first arch. Scales large, 125-143 in first row above lateral line.
SIZE: Length to 84 cm TL.
DISTRIBUTION: Point Hope, Alaska, to California. Anadromous and freshwater populations; at sea, stays close to shore for a while then moves into greater ocean.
(Kessler 1985, Mecklenburg et al. 2002)

Oncorhynchus tshawytscha

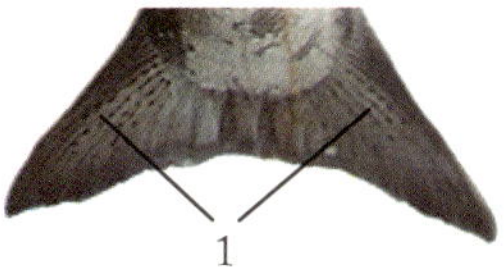

Oncorhynchus keta

Spawning male

Oncorhynchus nerka

Spawning female

HERRING

Pacific herring

Clupea pallasii

DESCRIPTION: Metallic blue-green dorsally; silvery ventrally; peritoneum black. Body moderately compressed. Abdominal scutes present, making a moderately developed ventral keel. Lower jaw slightly projects beyond upper (1); maxilla does not extend beyond eye. No enlarged scales on base of caudal fin; no adipose fin; no lateral line; small finlike structure at base of pelvic fin. Dorsal fin centered at about midpoint of body or slightly posterior to midpoint.

SIZE: Length to 46 cm TL.

DISTRIBUTION: Arctic Ocean to northern Baja California. Depth from surface to 250 m.

(Kessler 1985, Mecklenburg et al. 2002)

SMELTS

Rainbow smelt

Osmerus mordax

DESCRIPTION: Bluish to bluish-black or brown, with fine black speckling dorsally; silver-white laterally and ventrally. Mouth large, maxilla reaches to or beyond posterior margin of eye. Pelvic fin starts behind the origin of the dorsal fin; medium-sized scales. Vomer with one pair of large canines; palatine teeth larger anteriorly. Few gill rakers (4-6) on upper limb of first arch.

SIZE: Length to 25 cm TL.

DISTRIBUTION: Beaufort Sea to British Columbia. Anadromous; from surface to depth of 150 m.

(Kessler 1985, Mecklenburg et al. 2002)

Clupea pallasii

Osmerus mordax

Capelin

Mallotus villosus

DESCRIPTION: Blue or olive green dorsally; bright silvery-white laterally and ventrally; scales very small. Adipose fin flattened and attached for its entire length (1). Cucumber smell when fresh. Mouth moderate, maxilla extends to or almost to mid-eye.

SIZE: Length to 25 cm TL.

DISTRIBUTION: Beaufort Sea to Strait of Juan de Fuca; circumboreal in the Arctic Ocean. Pelagic, from surface to depth of 200 m.

(Kessler 1985, Mecklenburg et al. 2002)

Mallotus villosus

Eulachon

Thaleichthys pacificus

DESCRIPTION: Bluish to bluish-black or brown; silvery white laterally and ventrally. Adipose fin rounded, and attached only in front (1); pelvic fin starts in front of dorsal fin. Mouth large, maxilla reaches to or beyond posterior margin of eye; obvious concentric striae on operculum.

SIZE: Length to 25 cm TL.

DISTRIBUTION: Bering Sea, Aleutian Islands through coastal Gulf of Alaska to central California. Anadromous; from surface to depth of 300 m.

(Kessler 1985, Mecklenburg et al. 2002)

Thaleichthys pacificus

CODS

Pacific (gray) cod

Gadus macrocephalus

DESCRIPTION: Light gray-brown dorsally, whitish ventrally; brown to yellowish spots and blotches dorsolaterally; lateral line pale; median fin margins white; peritoneum black. Barbel (1) about as long as eye diameter; anus below second dorsal fin (2).
SIZE: Length to 99 cm TL.
DISTRIBUTION: Bering Strait to southern California. Benthic, also pelagic over deep water; near surface to depth of 875 m.
(Kessler 1985, Mecklenburg et al. 2002)

Gadus macrocephalus

Pacific tomcod

Microgadus proximus

DESCRIPTION: Grayish yellow or tan to olive green dorsally, without spots or mottling; white ventrally; fins gray to brown, tips dusky; peritoneum speckled black. First dorsal fin high, tip extends past origin of second dorsal fin when depressed (1), two anal fins; caudal fin straight; pectoral fin nearly reaches origin of anal fin. Anus below posterior fourth of first dorsal fin (2). Barbel short, about as long as pupil diameter (3).
SIZE: Length to 37 cm TL.
DISTRIBUTION: Southeastern Bering Sea and Aleutian Islands to central California. Demersal, at depths of 25-260 m over sand bottom.
(Kessler 1985, Mecklenburg et al. 2002)

Microgadus proximus

Saffron cod

Eleginus gracilis

DESCRIPTION: Body washed with yellow; brown to gray-green with mottling dorsally; yellow to white ventrally; dorsal and caudal fin margins white; pectoral fin yellow. Barbel about as long as pupil diameter (1). Anus under space between first and second dorsal fin (2), or under rear edge of first dorsal fin; first dorsal fin not high, tip barely reaches origin of second dorsal fin when depressed (3).
SIZE: Length to 55 cm TL.
DISTRIBUTION: Beaufort Sea and Chukchi Sea through eastern Bering Sea to southeastern Gulf of Alaska, and east in Arctic Ocean. Demersal, usually at depths less than 60 m.
(Kessler 1985, Mecklenburg et al. 2002)

Eleginus gracilis

Walleye pollock

Theragra chalcogramma

DESCRIPTION: Olive green to brown dorsally; silvery, with dark mottling and blotches dorsally and interrupted stripes of dark brassy olive laterally; white ventrally. Chin barbel minute or absent. Tail slightly forked or straight. Eye large; snout long, head profile flattened; teeth in upper jaw in a velvet-like band; 5-7 gill rakers on upper arch.
SIZE: Length to 91 cm TL.
DISTRIBUTION: Chukchi Sea through Bering Sea and Aleutian Islands to central California. Usually demersal at depths of 30-300 m.
(Kessler 1985, Mecklenburg et al. 2002)

Theragra chalcogramma

ROCKFISHES

The most useful characters for identifying rockfishes are coloration, head and cheek spines, fin spine and ray counts and relative size, and fin shape. Appearance of the symphyseal knob, size of the mouth, shape of the head between eyes (concave to convex), scale coverage, and gill raker numbers may also be helpful. Many species are very similar to others and both photographs and descriptions must be used to confirm species identification. There are 36 species of rockfish in Alaska—32 in the genus *Sebastes*, three in the genus *Sebastolobus*, and one in the genus *Adelosebastes*; 19 species are included here. *Guide to Northeast Pacific Rockfishes* by D.E. Kramer and V.M. O'Connell, *Pacific Fishes of Canada* by J.S. Hart, and *Guide to the Coastal Marine Fishes of California* by D.J. Miller and R.N. Lea may be helpful in identifying other rockfish.

EXTERNAL FEATURES OF ROCKFISHES

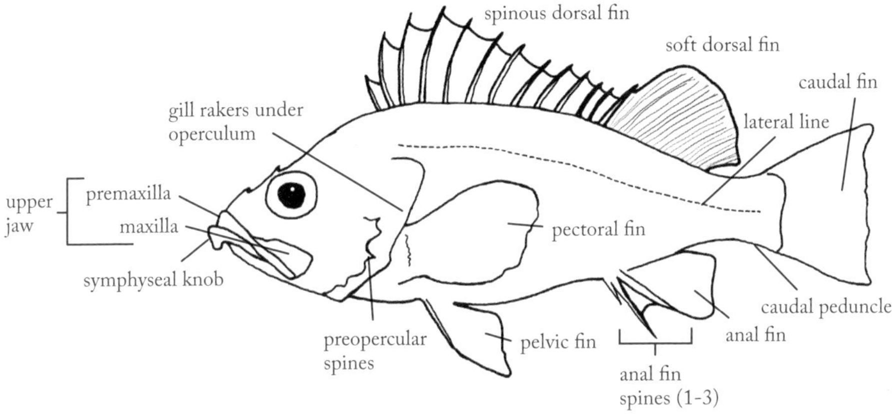

Rockfish Head Spines

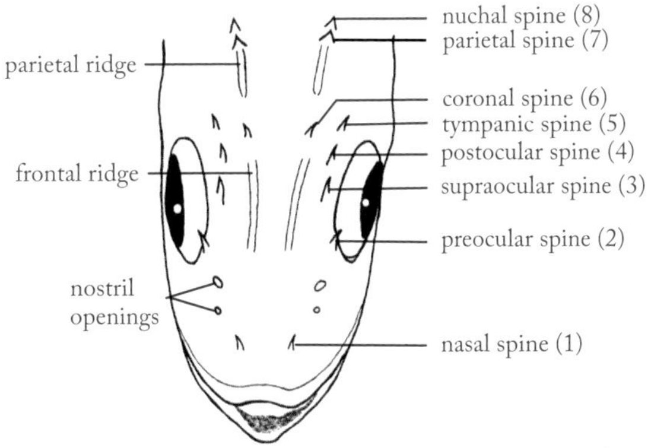

Shortspine thornyhead

Sebastolobus alascanus

DESCRIPTION: Bright red to pink; usually with dusky to black areas on fins; underside of head white; gill cavity mostly pale; peritoneum white, or white with black dots. Knifelike ridge below eye (1); 3 anal fin spines; 15-17 dorsal fin spines, third dorsal fin spine shorter than fourth or fifth dorsal fin spines; body round in cross section. Pectoral fin notched. The longspine thornyhead *S. altivelis* can be found in the Gulf of Alaska. It differs from the shortspine thornyhead by having the third dorsal fin spine longer than other dorsal fin spines.
SIZE: Length to 80 cm SL.
DISTRIBUTION: Bering Sea and Aleutian Islands to central Baja California. Demersal at depths of 17-1,524 m.
(Mecklenburg et al. 2002)

Yelloweye rockfish

Sebastes ruberrimus

DESCRIPTION: Eyes yellow; red dorsally, yellow-orange laterally, white to yellow ventrally; fins usually with black tips; juveniles and young adults darker, redder, with more black, and with white stripe along lateral line and shorter stripe below (1). Mouth and gill cavity white with some light pink or yellow. Rough ridges on top of head, head spines strong; 3 anal fin spines; 13 dorsal fin spines.
SIZE: Length to 91 cm TL.
DISTRIBUTION: Eastern Aleutian Islands south of Unalaska Island to southern California. In schools near surface to depth of 549 m.
(Mecklenburg et al. 2002)

Sebastolobus alascanus

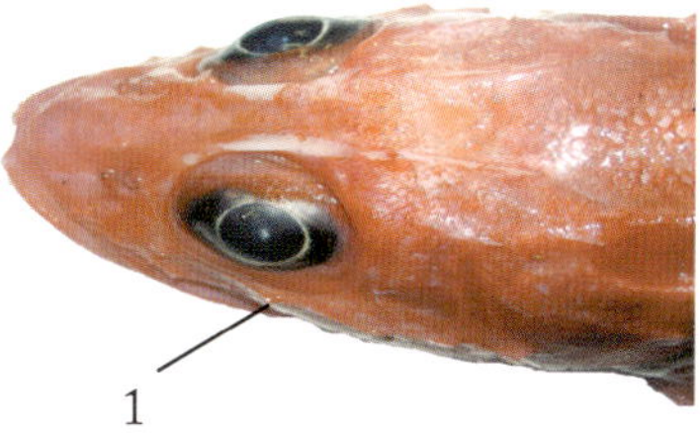

Juvenile

Sebastes ruberrimus

Shortraker rockfish

Sebastes borealis

DESCRIPTION: Pink to orange-pink body with vague dusky bands that are most prominent in younger fish; belly just as dark or only a little lighter than sides. Fins reddish, edged with black. Eye yellow; mouth, gill cavity, and gill membranes red to red-black. Large pores on lower jaw (1) which are much larger than other species of *Sebastes*. Head spines 1-5, 7, 8 moderate to strong. Maxilla extends to rear of pupil or to rear of eye. Anal fin spine 2 shorter than 3. The shortraker is distinguished from the rougheye rockfish *S. aleutianus* by having no spine, or rarely one weak spine, on rim below eye, shorter gill rakers on first arch (2), and pink or orange-pink coloration.
SIZE: Length to 120 cm TL.
DISTRIBUTION: Bering Sea at Navarin Canyon and Aleutian Islands to southern California. Near or on bottom in deep water offshore to depth of about 1,200 m.
(Mecklenburg et al. 2002, Kramer and O'Connell 2003)

Rougheye rockfish

Sebastes aleutianus

DESCRIPTION: Orange-red body, occasionally with dark blotches. All head spines usually present; 3 anal fin spines; 13 dorsal fin spines. Two or more spines on rim below eye (1). Fin and tail edges black (2). This widespread and commercially important rockfish has been considered a single variable species, with light and dark-colored forms. Genetic analysis has verified the presence of two species, *S. aleutianus* and *S. melanostictus*. *S. aleutianus* is pale, often has dark mottling on the dorsum in diffuse bands, and does not have distinct dark spots on the spinous dorsal fin.
SIZE: Length to 97 cm TL.
DISTRIBUTION: Eastern Aleutian Islands and southeastern Bering Sea to California. Near and on bottom at depths of 25-900 m. Overlaps depth distribution of *S. melanostictus* but typically shallower.
(Kessler 1985, Mecklenburg et al. 2002, Orr and Hawkins 2008)

Sebastes borealis

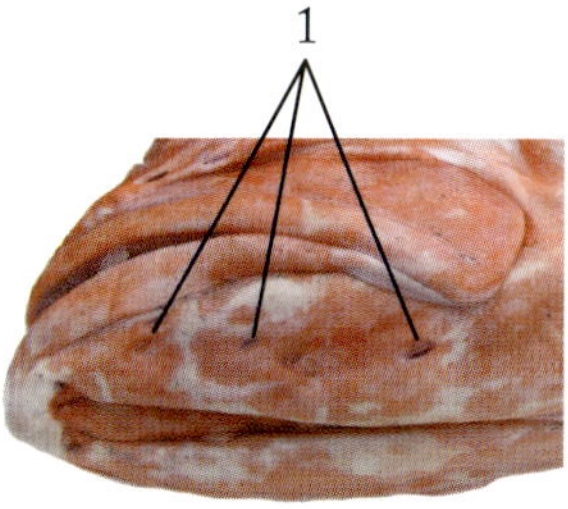

Sebastes aleutianus

Adult

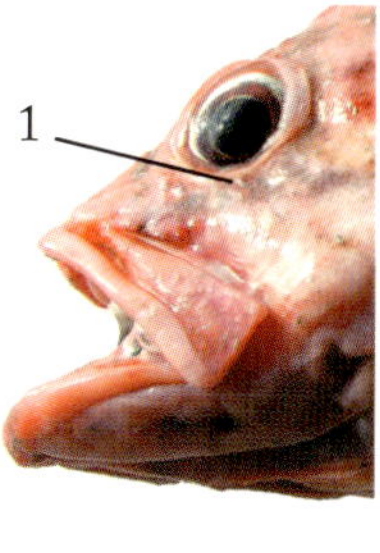

Blackspotted rockfish

Sebastes melanostictus

DESCRIPTION: Body color light to dark, often heavily mottled. Light specimens pink to red with greenish-black spotting below and above lateral line and on dorsum. Dark specimens with black to greenish wash over red to pink background, heavily mottled over entire body and spotted at and above the lateral line. Darker overall than *S. aleutianus* and spotting is nearly always present on the spinous dorsal fin. Longer dorsal fin, spines, and gill rakers than *S. aleutianus*. Other differences include number of gill rakers, lateral-line pores, and dorsal-fin rays.
SIZE: Length to 54 cm SL.
DISTRIBUTION: Central Japan, through the Aleutian Islands and Bering Sea, to southern California. Depth to at least 1,000 m. Commonly found at deeper depths than *S. aleutianus*.
(Orr and Hawkins 2008)

Pacific ocean perch

Sebastes alutus

DESCRIPTION: Bright to light red; olive brown blotches dorsally, and on caudal peduncle, base of fin, and operculum; fins red without white areas. Mouth mainly pink; gill cavity dusky; peritoneum gray with black dots or black. Prominent symphyseal knob on lower jaw (1), directed forward; maxilla extends to mid-eye or to rear edge of eye (2).
SIZE: Length to 55 cm TL.
DISTRIBUTION: Bering Sea at Navarin Basin and Aleutian Islands to central Baja California. Near surface to depth of 825 m.
(Mecklenburg et al. 2002)

Sebastes melanostictus

Sebastes alutus

Harlequin rockfish

Sebastes variegatus

DESCRIPTION: Reddish pink to deep red with olive or brown blotches; posterior ⅔ of lateral line in well-defined clear area (1); second dorsal, anal, and caudal fins black with red margins; clear red terminal band on caudal fin (2). Peritoneum dark brown to black in color. Slender and elongate compared to the sharpchin rockfish *S. zacentrus*. Head spines 1, 2, 4, 5, 7 moderate to strong. Symphyseal knob weak or absent; maxilla extends to mid-eye or to rear of pupil. Anal fin spine 2 longer and stronger than 3. The harlequin rockfish has the same head spination as *S. zacentrus*, but the second anal fin spine is longer than the third. It can also be differentiated by the missing prominent ridge on the bone below the nostril, and by an unpigmented band along the lateral line.
SIZE: Length to 37 cm TL.
DISTRIBUTION: Southeastern Bering Sea and Aleutian Islands to Pacific Ocean off Oregon. Deep inshore and offshore waters; adults at depths of 49-558 m; juveniles shallower.
(Mecklenburg et al. 2002)

Sharpchin rockfish

Sebastes zacentrus

DESCRIPTION: Pink to yellow pink, yellow pink on sides, light below. Five to 6 vague dark-brown blotches extend onto dorsal fin and below lateral line (1); reddish yellow ventrally; dark forked bar from eye to operculum. Mouth and gill cavity pink and white with yellow mottling; peritoneum black (2). Head spines 1,2,4,5,7 strong. Prominent ridge on bone below nostril (suborbital bone). Symphyseal knob strong, directed forward. Maxilla extends to mid-eye or to rear edge of pupil. Anal fin spine 2 slightly longer than anal fin spine 3.
SIZE: Length to 45 cm TL.
DISTRIBUTION: Attu Island, Aleutian Islands to southern California. Soft bottom and on seamounts at depths of 25-660 m.
(Mecklenburg et al. 2002, Kramer and O'Connell 2003)

OTHER RED ROCKFISHES with 13 dorsal fin spines. The darkblotched rockfish *Sebastes crameri* has a moderate knob directed downward, 5 dark blotches across the back, and all head spines except number 6. The redstripe rockfish *S. proriger* is distinguished by a red stripe along the lateral line and the missing number 3 and 8 head spines. The bocaccio *S. paucispinis* is distinguished by its long, projecting lower jaw, and a long upper jaw that extends past the eye. It has silvery-red sides and light brown, copper-brown, or red dorsal coloration.

Sebastes variegatus

Sebastes zacentrus

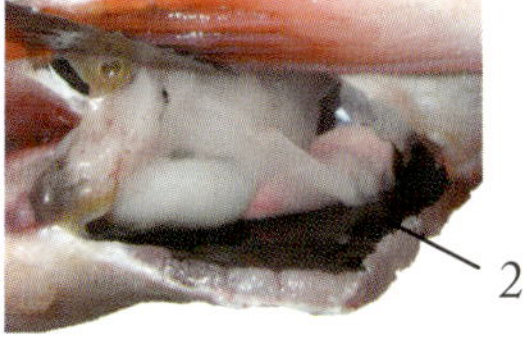

Northern rockfish

Sebastes polyspinis

DESCRIPTION: Dark greenish gray on red background with red-orange or brown specks and fine green spots; red-orange to white ventrally; dark bands radiate from eye (1); top of head dark; fins reddish to dusky red; light-colored band on lower side of the pectoral fins (2). Three anal fin spines; the northern is the only rockfish in Alaska with 14 dorsal fin spines. Gill cavity silver; peritoneum dark.
SIZE: Length to 41 cm TL.
DISTRIBUTION: Bering Sea at Pervenets Canyon and Aleutian Islands to northern British Columbia; western Bering Sea off northeastern Kamchatka and the Commander Islands. Mostly offshore at depths of 73-740 m.
(Mecklenburg et al. 2002)

Canary rockfish

Sebastes pinniger

DESCRIPTION: Gray mottled with yellow to orange, appearing orange overall; fins bright orange; 3 orange stripes across head; black blotch on posterior portion of spinous dorsal fin in fish under 30 cm. Lateral line in light gray area, contrasts with yellow-orange; mouth and gill cavity white, mottled with yellow, pink, and dusky. Head spines 1-5, 7 moderate to strong. Symphyseal knob weak; maxilla extends to rear of pupil or to rear edge of eye.
SIZE: Length to 76 cm TL.
DISTRIBUTION: Western Gulf of Alaska south of Shelikof Strait to northern Baja California. In schools around reefs and over hard bottoms; adults at depths of 18-425 m.
(Mecklenburg et al. 2002)

Sebastes polyspinis

Sebastes pinniger

Tiger rockfish

Sebastes nigrocinctus

DESCRIPTION: Color pink to orange, with 4 or 5 dark red or black bands; two dark bands radiate from eye (1); rear edge of opercular flap not covered by first body band (2). Interorbital space strongly concave, parietal ridges thick and coarse (3).
SIZE: Length to 61 cm TL.
DISTRIBUTION: Prince William Sound, Alaska, to California. Rocky reefs and boulder fields, at depths of 18-275 m.
(Kessler 1985, Mecklenburg et al. 2002, Kramer and O'Connell 2003)

Redbanded rockfish

Sebastes babcocki

DESCRIPTION: Body white or pink with 4 red bands; 1 or 2 dark bands radiate from eye (1); rear edge of opercular flap covered by first bar (2). Three anal spines; 13 dorsal fin spines. Maxilla extends to mid-eye or to rear of eye.
SIZE: Length to 65 cm TL.
DISTRIBUTION: Bering Sea at Zhemchug Canyon and Amchitka Island, Alaska, to southern California. Deepwater species on offshore reefs and seamounts at depths of 49-625 m.
(Mecklenburg et al. 2002)

China rockfish

Sebastes nebulosus

DESCRIPTION: Color blue-black, mottled with yellow, white, and bluish white splotches; broad yellow stripe starts at dorsal fin spines 3 and 4 and continues along lateral line to caudal fin (1). Head spines strong and high, encased in thick skin. Distal margin of anal fin rounded; caudal fin slightly rounded.
SIZE: Length to 45 cm TL.
DISTRIBUTION: Western Gulf of Alaska at Kodiak Island to southern California. Often found over reefs and crevices at depths of 3-128 m.
(Mecklenburg et al. 2002)

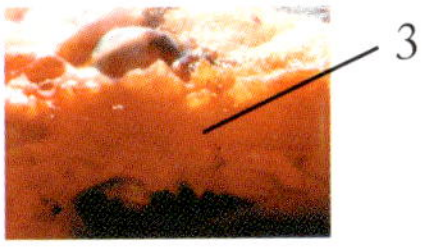

Sebastes nigrocinctus

Sebastes babcocki

Sebastes nebulosus

Black rockfish

Sebastes melanops

DESCRIPTION: Dark gray to black dorsally; lighter ventrally (orange and yellow blotching on head, tail, and pectoral and pelvic fins has been noted occasionally on specimens found in Kodiak). Spine on margin of opercular flap. Mouth and gill cavity light gray; peritoneum silvery white. Head spines weak to very weak; symphyseal knob absent.
SIZE: Length to 65 cm TL.
DISTRIBUTION: Aleutian Islands to southern California. School over rocky reefs from surface to depth of 366 m, usually found shallower than 150 m.
(Mecklenburg et al. 2002)

Dark rockfish

Sebastes ciliatus

DESCRIPTION: Uniformly dark blue to black, gradually lightening ventrally; peritoneum black. Smaller symphyseal knob (1), and fewer lateral line pores than *S. variabilis*. The dusky rockfish *S. variabilis* has been considered a single variable species with light and dark forms distributed in deep and shallow water, respectively. From morphological evidence two species corresponding with these color forms are now recognized. *S. ciliatus* is the dark shallow-water species and *S. variabilis* is the deeper water species.
SIZE: Length to 41 cm TL.
DISTRIBUTION: Western Aleutian Islands and eastern Bering Sea to British Columbia. Shallow water species found in depths of 5-160 m.
(Orr and Blackburn 2004)

Dusky rockfish

Sebastes variabilis

DESCRIPTION: Variable in color, ranging from greenish to yellowish brown dorsally, pink or lighter ventrally, olive or brown bands radiate from eye; fins tinged with pink; peritoneum black. Symphyseal knob on lower jaw moderate to large.
SIZE: Length to 41 cm TL.
DISTRIBUTION: Aleutian Islands and eastern Bering Sea to Oregon. Deepwater species at 12-675 m.
(Orr and Blackburn 2004)

Rare color variation

Sebastes melanops

Sebastes ciliatus

Sebastes variabilis

Silvergray rockfish
Sebastes brevispinis

DESCRIPTION: Dark greenish to silvery gray dorsally and laterally, pink to white ventrally; lateral line in narrow pinkish stripe; lower portions of anal, pectoral, and pelvic fins tinged orange or pink; lips blackish. Head spines 1, 2, 7 weak. Lower jaw strongly projects, symphyseal knob large. Maxilla extends to rear edge of pupil to rear edge of eye; interorbital space convex. Anal fin spine 2 slightly shorter or about same length as anal fin spine 3.
SIZE: Length to 71 cm TL.
DISTRIBUTION: Southeastern Bering Sea to Pacific Ocean off central Baja California. Around reefs and over soft bottoms, from surface to 375 m.
(Mecklenburg et al. 2002)

Yellowtail rockfish
Sebastes flavidus

DESCRIPTION: Olive green to grayish brown in color with fins tinged with yellow; lower rays of pectoral fin tinged with pink; mouth and gill cavity light gray; peritoneum silvery white. Prominent symphyseal knob on lower jaw.
SIZE: Length to 66 cm TL.
DISTRIBUTION: Eastern Aleutian Islands south of Unalaska Island to southern California. Schooling rockfish, surface to 549 m.
(Kessler 1985, Mecklenburg et al. 2002)

Quillback rockfish
Sebastes maliger

DESCRIPTION: Grayish brown mottled with yellow anteriorly, and orange-brown spots on lower, anterior part of body; fins dark brown or blackish except for large white to yellow area on spinous dorsal fin. Dorsal fin deeply incised and high. Mouth and gill cavity white; peritoneum silvery white.
SIZE: Length to 61 cm TL.
DISTRIBUTION: Western Gulf of Alaska at Kodiak Island to southern California. Found on rocky bottoms and reefs inshore to depth of 274 m.
(Mecklenburg et al. 2002)

Sebastes brevispinis

Sebastes flavidus

Sebastes maliger

SABLEFISH

Sablefish (blackcod)

Anoplopoma fimbria

DESCRIPTION: Small fish are blue-green or copper-brown; large fish are black or dark gray. Two well-separated dorsal fins; 1 anal fin with origin below origin of second dorsal fin. Uniformly scaled, skin like fine sandpaper. Streamlined shape.
SIZE: Length to 114 cm TL.
DISTRIBUTION: Bering Sea south of St. Lawrence Island and Aleutian Islands to central Baja California; western Bering Sea and Commander Islands. Near bottom over soft substrate at depths to 2,740 m.
(Kessler 1985, Mecklenburg et al. 2002)

GREENLINGS

Atka mackerel

Pleurogrammus monopterygius

DESCRIPTION: Yellowish gray to olive green; dorsal fin yellow or orange; lower part of pectoral fin black (1); light and dark vertical bars. Males bright yellow during summer breeding months. Long-based dorsal fin without a notch (2); forked tail; 5 lateral lines; no anal fin spine.
SIZE: Length to 60 cm SL.
DISTRIBUTION: Bering Sea and Aleutian Islands to southern California; western Bering Sea and Commander Islands. Lower intertidal to depth of 575 m.
(Kessler 1985, Mecklenburg et al. 2002)

Lingcod

Ophiodon elongatus

DESCRIPTION: Color variable, usually with dark-brown or copper blotches arranged in clusters. Large mouth with sharp teeth (1). Long-based dorsal fin with a notch; prominent cirri on head (2); 1 lateral line.
SIZE: Length to 152 cm TL.
DISTRIBUTION: Southwestern Gulf of Alaska off Shumagin Islands to northern Baja California. Near and on bottom around rocky areas. Inshore and to depth of 475 m.
(Kessler 1985, Mecklenburg et al. 2002)

Anoplopoma fimbria

Pleurogrammus monopterygius

Ophiodon elongatus

Kelp greenling

Hexagrammos decagrammus

DESCRIPTION: Male olive brown to bluish gray, with irregular blue spots mostly anteriorly. Female background color lighter, speckled all over with reddish brown to gold spots, face and fins with various shades of golden yellow; eye yellow. Two pairs of cirri on head, one above eyes and one smaller pair on occiput in slight depression. Dorsal fin divided at about middle by deep notch; usually 1 small spine at anterior end of anal fin. Five lateral lines; fourth lateral line long, extends nearly to or past end of anal fin.
SIZE: Length to 53 cm TL.
DISTRIBUTION: Aleutian Islands to Gulf of Alaska coasts to southern California. Rocky reefs and kelp beds to depth of 46 m.
(Mecklenburg et al. 2002)

Rock greenling

Hexagrammos lagocephalus

DESCRIPTION: Color variable, often green to brown with dark (red or orange in male) mottling; lighter ventrally; often with dark spot above pectoral fin base; mouth cavity typically blue, flesh and bones sometimes blue. Female often solid mahogany colored or with oblong, blue or other light-colored spots on body; eye red. Large cirrus above each eye (1). Dorsal fin divided at about middle by deep notch. Five lateral lines.
SIZE: Length to 61 cm TL.
DISTRIBUTION: Eastern Bering Sea and Aleutian Islands to California; western Bering Sea and Commander Islands. Shallow rocky areas, especially along exposed coasts, to depth of 80 m.
(Mecklenburg et al. 2002)

Male

Hexagrammos decagrammus

Female

Male

Hexagrammos lagocephalus

Female

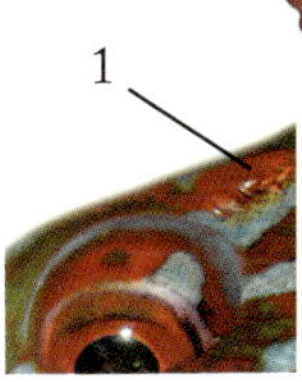

Masked greenling

Hexagrammos octogrammus

DESCRIPTION: Green to brown; dark mottling and spots; eye red. Caudal peduncle stout, depth greater than snout length. Maxilla extending to front of pupil. One moderately large cirrus above each eye. Dorsal fin divided at about middle by a notch; dorsal fin spines usually fewer than 20, with short cirri at tips. Five lateral lines.
SIZE: Length to 28 cm TL.
DISTRIBUTION: Bering Sea and Aleutian Islands to northern British Columbia. Western Bering Sea to Okhotsk and Japan seas. Nearshore areas at depths of 6-31 m.
(Mecklenburg et al. 2002)

Hexagrammos octogrammus

OTHER GREENLINGS. The whitespotted greenling *H. stelleri* is distinguished from other greenlings by the presence of small white spots on the head and body and a slender caudal peduncle.

SCULPINS

The sculpins described in this guide are in four related families, most of which are in the cottid group. Mecklenburg et al. 2002 note that there are 87 cottid species in 35 genera, nine fathead sculpins, seven sailfin sculpins, and one grunt sculpin found in Alaska waters. Sculpins make up a large, highly variable group of fishes, differing from their closest look-alikes (rockfishes) in that the sculpin body is never completely covered by rows of regular scales. Most sculpin bodies taper gradually toward the posterior half of the body, whereas rockfish are generally thick-bodied. Also, the heads of some sculpins appear rather squat and proportionally larger than those of rockfishes. As with rockfishes, key features for sculpin identification include color, head and cheek spines, and dorsal fin configuration.

Grunt sculpin

Rhamphocottus richardsonii

DESCRIPTION: Creamy yellow with irregular dark-brown streaks running obliquely across body and radiating from eye (1); caudal peduncle and fins orange to red color. Head large; snout elongate (2); dorsal profile high. Lower pectoral fin rays free. Head and body covered with multipointed spinous plates.
SIZE: Length to 9 cm TL.
DISTRIBUTION: Gulf of Alaska to southern California. Found on rocky and rubble sand mixed bottoms from the intertidal zone to depth of 165 m.
(Mecklenburg et al. 2002)

Rhamphocottus richardsonii

Roughspine sculpin

Triglops macellus

DESCRIPTION: Dark brown to olive green dorsally, cream ventrally; dark saddles across back. Dark, narrow bars on rays of dorsal and caudal fins and upper part of pectoral fin (1). Caudal fin slightly indented to nearly straight; fin membranes clear. Head and body depressed; body very narrow and elongate; snout pointed. Body above lateral line densely covered with fine prickly scales, skin below lateral line folded, with serrated edges.
SIZE: Length to 30 cm TL.
DISTRIBUTION: Eastern Bering Sea from north of St. Matthew Island and Aleutian Islands to eastern Pacific Ocean off Washington. Depths of 18-275 m.
(Mecklenburg et al. 2002)

Scissortail sculpin

Triglops forficatus

DESCRIPTION: Dark brown dorsally, cream ventrally; usually with dark saddles. Tail deeply forked, with black tips (1); black bar and silver patch in front of pectoral fin (2). Body above lateral line densely covered with fine prickly scales. Skin below lateral line folded, with serrated edges.
SIZE: Length to 32 cm TL.
DISTRIBUTION: Bering Sea, Aleutian Islands, and eastern Gulf of Alaska. Found on various bottom types at depths of 20-425 m.
(Kessler 1985, Mecklenburg et al. 2002)

THE RIBBED SCULPIN *Triglops pingelii* is also widespread in Alaska waters, but can be distinguished from the roughspine and scissortail sculpins as it has dark spots or streaks below the lateral line that are either continuous (males) or broken (females).

Red Irish lord

Hemilepidotus hemilepidotus

DESCRIPTION: Color variable, combinations of solid or mottled red to orange, pink or white, brown, and black mottling; whitish ventrally. Two bands of scales obvious (1); 4 or 5 rows of scales in dorsal band; 6 or 7 rows in ventral band. Dorsal fin continuous; first 3 dorsal spines shorter than fourth; 4 preopercular spines; nasal cirri present.
SIZE: Length to 51 cm TL.
DISTRIBUTION: Southeastern Bering Sea, Aleutian Islands to California, western Bering Sea and Commander Islands to Pacific Ocean. Tide pools and to depths of 48 m.
(Kessler 1985, Mecklenburg et al. 2002)

Triglops macellus

Triglops forficatus

Hemilepidotus hemilepidotus

Yellow Irish lord

Hemilepidotus jordani

DESCRIPTION: Yellowish tan to dark brown; vague dark saddles. Two bands of scales obvious (1); white chin (2); yellow patch in front of pectoral fin and on tail (sometimes missing) (3); 4 or 5 rows of scales in band below dorsal fin. Dorsal fin continuous; third spine shorter than fourth, forms notch in profile.

SIZE: Length to 50 cm TL.

DISTRIBUTION: Southeastern Chukchi Sea and Bering Sea to Aleutian Islands and southeastern Alaska. Found on soft bottoms, usually at subtidal depths to 110 m. (Kessler 1985, Mecklenburg et al. 2002)

OTHER IRISH LORDS. The brown Irish lord *Hemilepidotus spinosus* has been found in the northern Gulf of Alaska but seems to be more a southern species. It has 6 to 8 rows of scales in a band under the dorsal fin. The longfin Irish lord *H. zapus* is found along the Aleutian Islands and in the southwestern Gulf of Alaska. The narrow width between the eyes is a distinguishing characteristic. The butterfly sculpin *H. papilio*, is common in the northern Bering Sea though not commonly named an Irish lord.

Threaded sculpin

Gymnocanthus pistilliger

DESCRIPTION: Brown dorsally, reticulated with light brown; white ventrally; 4 vague, dusky saddles. First dorsal fin of males with white spots; dorsal, caudal, pectoral, and sometimes pelvic fins barred. Tips of pistil-like cirri of males bright white. Four preopercular spines. First dorsal fin high and sail-like (1), more pronounced in males; bony bumps between head and first dorsal fin (2); dorsal fin spines usually 9-10, dorsal rays 14-15, anal rays 16, pectoral 18-19; total count for all fins 80 or less. Space between eyes smooth.
SIZE: Length to 29 cm TL.
DISTRIBUTION: Bering Sea, Aleutian Islands, and southeastern Alaska. Soft bottoms in shallow water, rarely deeper than 100 m.
(Kessler 1985, Mecklenburg et al. 2002)

Gymnocanthus pistilliger

Armorhead sculpin
Gymnocanthus galeatus

DESCRIPTION: Brown to olivaceous or gray, lightly reticulated dorsally; white ventrally (1); 4 or 5 blackish saddles extend below the lateral line; dorsal, caudal and pectoral fins barred (2). Head slightly compressed; interorbital space narrow. Four preopercular spines (3); the first is elongate, bearing 1-3 spinules on dorsal surface. Dorsal fin rays 15-17, anal 18 or 19, pectoral 20. Rough plates cover top of head and interorbital space; lateral line scales as embedded tubules.
SIZE: Length to 36 cm TL.
DISTRIBUTION: Bering Sea and Aleutian Islands to central British Columbia; western Bering Sea and Commander Islands to northern Sea of Japan. Soft bottom near shore and to depth of 579 m.
(Mecklenburg et al. 2002)

Leister sculpin
Enophrys lucasi

DESCRIPTION: Background color light olive, thickly covered with dusky spots; 3 or 4 dark bands; some cirri centered in pale spots on side; pale yellow ventrally. Fins with pale yellow bars and spots; anal and pelvic fins with dark spots along rays. Body robust anteriorly; head massive; maxilla extends beyond pupil. Eye orbit protrudes above profile of head. Preopercular spine shorter than the distance from the eye to the end of the hump (1); forked spine above mouth bluntly rounded (2), extends only partly over upper lip. Lateral line plates with 1 or 2 keels (3).
SIZE: Length to 25 cm TL.
DISTRIBUTION: Bering Strait off Little Diomede Island to British Columbia. Shallow water near shore to depth of about 17 m.
(Kessler 1985, Mecklenburg et al. 2002)

Brightbelly sculpin
Microcottus sellaris

DESCRIPTION: Body purplish, spotted and mottled in black and white, or red and black, with 2 or more white bars and bright yellow belly (1). Pelvic fins connected to belly by a membrane for about half their length, with black or red spots along rays; female not as brightly colored. Head deep, profile abrupt.
SIZE: Length to 13 cm TL.
DISTRIBUTION: Chukchi Sea to southern Bering Sea and Aleutian Islands. Coastal areas to depth of about 50 m.
(Kessler 1985, Mecklenburg et al. 2002)

Gymnocanthus galeatus

Enophrys lucasi

Microcottus sellaris

Shorthorn sculpin

Myoxocephalus scorpius

DESCRIPTION: Greenish brown to blackish, with yellow orange tinges and dark mottling; second dorsal and anal fins with light and dark vertical bars; first and second dorsal fin barely separated (1); belly white. Head depressed; top of head strongly verrucose or not, protuberances or spines on each side; usually 3 preopercular spines. Platelike scales above lateral line with depressed centers (2); fewer scattered plates below lateral line. Highly variable species.
SIZE: Length to 60 cm, usually less than 35 cm TL.
DISTRIBUTION: Beaufort Sea to northern British Columbia; Laptev Sea to Kamchatka Bay and Commander Islands. Shallow water near shore to depth of 550 m.
(Kessler 1985, Mecklenburg et al. 2002)

Plain sculpin

Myoxocephalus jaok

DESCRIPTION: Gray, with many small black spots; white ventrally; no saddles or bands on body; large white spots ventrolaterally; fins barred and spotted. First preopercular spine twice as long as second. Well-separated dorsal fins (1).
SIZE: Length to 61 cm TL.
DISTRIBUTION: Chukchi Sea to Bering Sea, eastern Aleutian Islands, and Gulf of Alaska. Sand and mud bottoms from intertidal zone to 80 m.
(Kessler 1985, Mecklenburg et al. 2002)

Great sculpin

Myoxocephalus polyacanthocephalus

DESCRIPTION: Olive brown or gray dorsally, mottled with yellow to orange-red; usually with 3 or 4 dark, irregular saddles (1); belly white; small white spots ventrolaterally; fins barred and spotted. Well-separated dorsal fins (2). Head wide; top and sides of head covered in warts. First preopercular spine twice as long as second (3); cluster of short ridges behind eye. Lateral line with 3 rows of pores.
SIZE: Length to 76 cm TL.
DISTRIBUTION: Chukchi Sea to Bering Sea, eastern Aleutian Islands and Gulf of Alaska to Kachemak Bay; southeastern Alaska at Limestone Inlet; to Okhotsk and Japan seas. Sand and mud bottoms and around rocks, usually from intertidal zone to depths less than 200 m.
(Mecklenburg et al. 2002)

Myoxocephalus scorpius

Myoxocephalus jaok

Myoxocephalus polyacanthocephalus

Preopercular spines

OTHER SPECIES OF *MYOXOCEPHALUS*. The fourhorn sculpin *M. quadricornis* has comblike crests on its head. The arctic sculpin *M. scorpioides* has wartlike head protuberances, confluent dorsal fins, and 15 pectoral fin rays. The warthead sculpin *M. niger* is dark black and has wartlike head protuberances with long feather-like cirri. The frog sculpin *M. stelleri* has greatly thickened lips and the top of the head is verrucose.

Bigmouth sculpin

Hemitripterus bolini

DESCRIPTION: Mottled gray to brown dorsally, lighter ventrally; 4 or 5 vague darker saddles across back; reddish, yellow-brown, and orange patches and highlights. Head large and depressed; mouth extremely large, maxilla extends well beyond eye; numerous cirri, many complex, mainly on head (1); 13-15 free dorsal fin spines each with a flag-like skin pennant (2).

SIZE: Length to 69 cm TL.

DISTRIBUTION: Bering Sea from Cape Navarin to north side of Alaska Peninsula, and Aleutian Islands to northern California. Bottom at depths of 25-925 m.

(Kessler 1985, Mecklenburg et al. 2002)

Hemitripterus bolini

Crested sculpin

Blepsias bilobus

DESCRIPTION: Orange-brown to olive green with darker blotches, body compressed; 4 preopercular spines. Prominent, simple cirri on snout and lower jaw. Spinous dorsal fin base shorter than soft dorsal; spinous dorsal fin without deep notch; pectoral fin rays 15 or more. Prickles densely cover body.
SIZE: Length to 27 cm TL.
DISTRIBUTION: Chukchi and Bering seas and Aleutian Islands to northern British Columbia; western Chukchi Sea to Commander Islands. Near bottom to depth of 190 m.
(Mecklenburg et al. 2002)

Blepsias bilobus

Silverspotted sculpin

Blepsias cirrhosus

DESCRIPTION: Golden brown to olive green dorsally, copper red to yellow ventrally; bright silvery white patches below lateral line and often on fins and cheeks (1). Prominent cirri on snout and lower jaw. Spinous dorsal fin base shorter than soft dorsal; spinous dorsal fin deeply notched, nearly forming two fins (2); pectoral fin rays 14 or fewer.
SIZE: Length to 20 cm TL.
DISTRIBUTION: Southeastern Bering Sea and Aleutian Islands to central California. Near shore to depth of about 47 m.
(Mecklenburg et al. 2002)

Sailfin sculpin

Nautichthys oculofasciatus

DESCRIPTION: Orange-brown to pale gray with blackish blotches; black bar through eye and across cheek (1). Sail-like first dorsal fin (2). Head spines strong, rounded; nasal spines well developed, sharp. Large cirrus on eye.
SIZE: Length to 20 cm TL.
DISTRIBUTION: Western Gulf of Alaska to southern California. Intertidal zone, sometimes in tide pools, and to depth of 110 m.
(Mecklenburg et al. 2002)

Spinyhead sculpin

Dasycottus setiger

DESCRIPTION: Pinkish gray with gray to dark-brown blotches and bars on head, body, and fins. Head large with a pair of prominent, blunt, skin-covered spines about halfway between eyes and first dorsal fin (1). Lower jaw protrudes; snout broad and humped. Four spines, variously directed above eye. Numerous cirri scattered on head and body. Dorsal fins separate or slightly connected (2).
SIZE: Length to 45 cm TL.
DISTRIBUTION: Bering Sea from Navarin Canyon and Aleutian Islands to Washington; western Bering Sea to Commander Islands. Soft bottom at depths of 15-850 m.
(Kessler 1985, Mecklenburg et al. 2002)

Blepsias cirrhosus

Nautichthys oculofasciatus

Dasycottus setiger

Darkfin sculpin

Malacocottus zonurus

DESCRIPTION: Light grayish brown, with vague dark-brown to blackish bands; fins with dark and light bands and white margins; caudal fin with black bands. Maxilla extends to mid-eye. Four preopercular spines, almost always with an accessory (fifth) spine in front of the second and directed laterally. Dorsal fins separate but touching or nearly touching; bases of fins covered with loose skin.
SIZE: Length to 35 cm TL.
DISTRIBUTION: Bering Sea and Aleutian Islands to Washington. Bottom at depths of 95-1,980 m.
(Mecklenburg et al. 2002)

Malacocottus zonurus

OTHER SCULPINS. For complete information on the 87 species documented in Alaska, see Mecklenburg et al. 2002.

POACHERS

Sawback poacher

Leptagonus frenatus

DESCRIPTION: Brown dorsally, whitish ventrally; dark blotches and bands; dorsal fins brown with white spots; caudal fin brown; pectoral fin brown, with or without white blotch on upper portion near base; pelvic fin white or brown. Body covered with bony plates, keel of body plates strongly developed, with sharp recurved spines (1). Mouth ventral and directed slightly forward; white barbels (2). (Also known as *Sarritor frenatus*.)

SIZE: Length to 24 cm SL.

DISTRIBUTION: Norton Sound, Bering Sea to northern British Columbia; western Bering Sea to Japan. On bottom at depths of 18-975 m.

(Kessler 1985, Mecklenburg et al. 2002)

Leptagonus frenatus

Fourhorn poacher

Hypsagonus quadricornis

DESCRIPTION: Reddish brown, with yellow, red, and brown bands and blotches; dark band along caudal fin margin (1). Body rises steeply behind head; barbel at tip of snout (2); high-set eyes; often has a fungus-like growth on its sides. Dorsal fins close together, anal fin origin below first dorsal fin.
SIZE: Length to 11 cm TL.
DISTRIBUTION: Bering Sea and Aleutian Islands to Washington. Rocky sand, gravel, or pebble bottoms at depths of 15-452 m.
(Mecklenburg et al. 2002)

Sturgeon poacher

Podothecus accipenserinus

DESCRIPTION: Yellowish brown to gray-brown with dark bands dorsally, white to orange ventrally. Body covered with bony plates. Dorsal and caudal fin margins black. Mouth ventral with gap when closed; yellow barbels (1).
SIZE: Length to 30 cm TL.
DISTRIBUTION: Bering Sea and Aleutian Islands to northern California; western Bering Sea to Commander Islands. Soft bottoms at depths of 2-300 m.
(Mecklenburg et al. 2002)

Blackfin poacher

Bathyagonus nigripinnis

DESCRIPTION: Brown body; bluish-black gill membrane; fins uniformly black or bluish-black. Lower jaw projects beyond upper jaw (1). Anal fin origin below dorsal fin interspace.
SIZE: Length to 24 cm TL.
DISTRIBUTION: Eastern Bering Sea and Aleutian Islands to northern California; western Bering Sea from Navarin Canyon to Commander Islands. Fine sand or mud bottoms at depths of 18-1,247 m.
(Mecklenburg et al. 2002)

Hypsagonus quadricornis

Podothecus accipenserinus

Bathyagonus nigripinnis

Alligatorfish

Aspidophoroides monopterygius

DESCRIPTION: Brownish to olive green dorsally, whitish ventrally; faint dark bands on body; pectoral fin with whitish lower lobe; caudal fin margin pale; pelvic fin white. Body slender, depth less than 12% standard length; covered with bony plates. One dorsal fin (1); 4-6 anal fin rays; 9-10 pectoral fin rays. Mouth terminal. Prominent nasal spines (2).
SIZE: Length to 18 cm TL.
DISTRIBUTION: Eastern Chukchi Sea, Bering Sea, and Aleutian Islands west to Amchitka Island, and through the Gulf of Alaska to Prince William Sound. Sand and mud bottoms at depths of 8-500 m.
(Kessler 1985, Mecklenburg et al. 2002)

LUMPSUCKER

Smooth lumpsucker

Aptocyclus ventricosus

DESCRIPTION: Dark gray, lighter ventrally. Globe-shaped body with smooth skin when inflated; skin naked, no tubercles or scales. Ventral sucking disk. First dorsal fin completely embedded under skin; pectoral fins almost touch each other at base (1).
SIZE: Length to 42 cm TL.
DISTRIBUTION: Bering Sea to Pacific Ocean south of Aleutian Islands and Gulf of Alaska to British Columbia. Pelagic, usually found in deep waters, taken from surface to depths greater than 500 m.
(Kessler 1985, Mecklenburg et al. 2002)

Aspidophoroides monopterygius

Aptocyclus ventricosus

SNAILFISHES

Liparis species have two nostrils on each side, pseudobranchs, a pelvic disk, trilobed teeth, and pectoral fin rays in about the same number or more than anal fin rays. The anus is usually about midway between the disk and the anal fin or closer to the anal fin than the disk. Body coloration variable within species. Most species of the genus inhabit coastal regions; they are typically found in tide pools, but are represented to a depth of about 450 m.

Variegated snailfish

Liparis gibbus

DESCRIPTION: Coloration highly variable: pink, brown with tan stripes, tan with fin margins black, or brownish with large blackish spots; peritoneum pale. Body elongate; ventral sucking disk (1); skin loose. Thumbtack prickles present in mature males on head, body, and dorsal fin (2). Dorsal fin not notched; dorsal and anal fins usually extend well onto caudal fin, anal fin slightly farther than dorsal (3); dorsal fin rays usually 41-44, anal fin rays usually 33-36.

SIZE: Length to 52 cm TL.

DISTRIBUTION: Arctic Alaska to Bering Sea and Aleutian Islands, and Gulf of Alaska to northern British Columbia. Benthic, usually at depths less than 200 m.

(Mecklenburg et al. 2002)

Liparis gibbus

Careproctus species have a single nostril on each side, no pseudobranchs, a pelvic disk, snout without barbels, and trilobed or simple teeth. Pectoral fin rays are typically fewer than anal fin rays. The anus is closer to the disk than to the anal fin. Body is without spots, stripes, or mottling, and any significant pigmentation is usually restricted to the fins. New species of *Careproctus* with variegated red, white, and gray color variations have been documented from the Aleutian Islands. Most *Careproctus* species inhabit dimly lighted regions at 100-3,335 m depth. The blacktail snailfish *C. melanurus* is common in the Bering Sea and Aleutian Islands but is not pictured here.

Salmon snailfish

Careproctus rastrinus

DESCRIPTION: Bright yellow pink to salmon color; skin transparent; peritoneum pale; stomach pale to dusky. Profile ascends steeply from blunt snout; body deep. Anterior dorsal and anal fin rays buried in gelatinous tissue; pectoral fins deeply notched, lower rays typically longer than upper and partly free. Nostril tube short but prominent. Disk 11-20% of head length, triangular (1). Anus close behind disk (2).
SIZE: Length to 49 cm SL.
DISTRIBUTION: Bering Sea and Aleutian Islands to southeastern Alaska. Soft sand, silt, and mud bottoms at depths of 55-913 m.
(Mecklenburg et al. 2002)

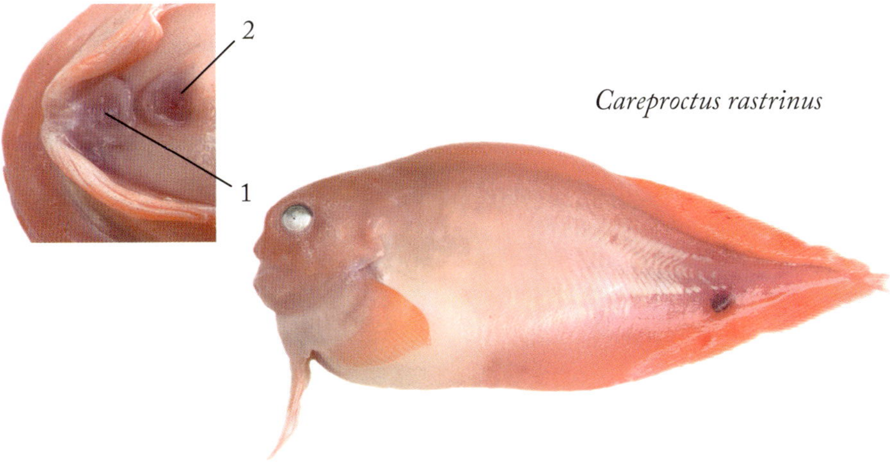

Careproctus rastrinus

Spectral snailfish
Careproctus phasma

DESCRIPTION: Pale pink color; body covered with small black dots; mouth, gill cavity, and peritoneum white or silvery. Skin opaque; body heavy anteriorly, short, tapering rapidly to caudal fin. Eye about 25% of head length. Disk larger than eye; anus close behind disk.
SIZE: Length to 10 cm TL.
DISTRIBUTION: Eastern Bering Sea, Amchitka Island. Mud, sand, gravel, and shell bottoms at depths of 84-504 m.
(Kessler 1985, Mecklenburg et al. 2002)

Blotched snailfish
Crystallichthys cyclospilus

DESCRIPTION: Body yellowish or pinkish; ovate or round, pink, orange, or red blotches on head, body, and fins; some blotches surrounded by darker border. Snout broadly rounded. Anterior dorsal and anal fin rays buried; pectoral fins notched, lower rays thick and tips free. *Crystallichthys* species are similar to *Careproctus* species but differ in having spots or bars on the body and a unique, horizontally divided pupil.
SIZE: Length to 29 cm TL.
DISTRIBUTION: Bering Sea and Aleutian Islands to Gulf of Alaska south of Shumagin Islands. Mud, sand, and gravel bottoms at depths of 54-265 m.
(Mecklenburg et al. 2002)

Combed snailfish
Allocareproctus kallaion

DESCRIPTION: Color light red, with darker red blotches scattered over body, dark speckling at origin of dorsal fin, anterior part of body blue-gray, posterior part of body blotched red; peritoneum black. Head large, mouth small, maxilla extends to anterior portion of orbit. Papillae present on many pores. Gill opening small, entirely above pectoral fin (1); gill raker count on first arch 14-18.
SIZE: Length to 20 cm SL.
DISTRIBUTION: Eastern to central Aleutian Islands. Depths of 278-458 m.
(Orr and Busby 2006)

Careproctus phasma

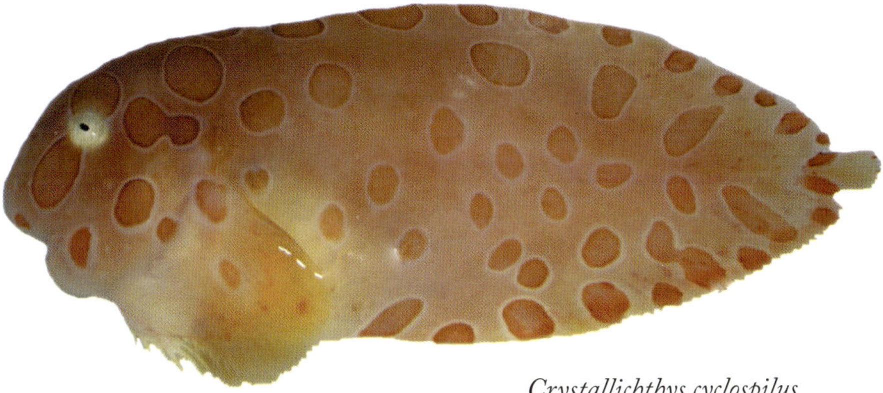

Crystallichthys cyclospilus

Allocareproctus kallaion

Cherry snailfish

Allocareproctus jordani

DESCRIPTION: Body red; iris silver gray (1). Teeth simple or with weak shoulders in a band of 7-11 oblique rows; mouth small; peritoneum black. A pigmented papilla on nasal pore helps distinguish this *Allocareproctus* from similar species. Gill opening small (2), entirely above pectoral fin or extends to pectoral-fin ray 2; gill raker count on first arch 8-12, short and blunt. Pelvic disk large; anus nearer disk than anal fin (3).
SIZE: Length to 17 cm SL.
DISTRIBUTION: Aleutian Islands. Depths of 75-631 m.
(Orr and Busby 2006)

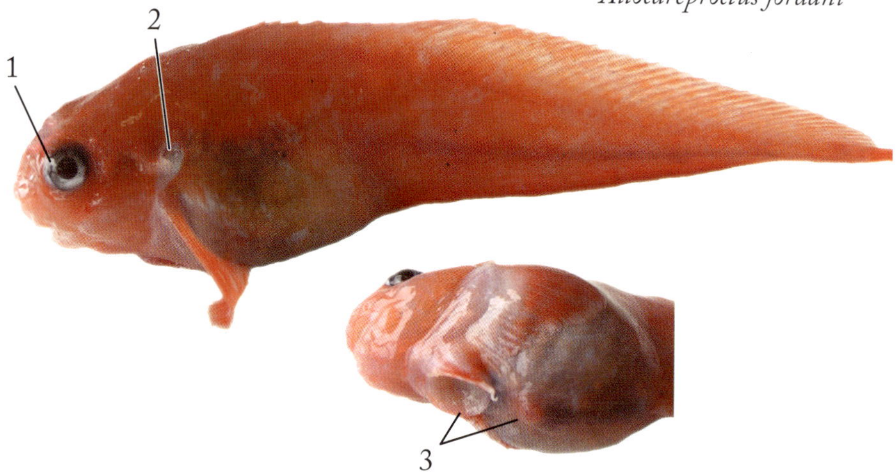

Allocareproctus jordani

OTHER SNAILFISHES. Many species of snailfish occur in Alaska. Mecklenburg et al. (2002) reference 17 species in the genus *Liparis*, 23 species in the genus *Careproctus*, and 18 other species from 8 genera. Identifications are difficult and not many snailfish species have been well documented. Specimens with questionable identifications should be preserved and retained for positive identification.

ARMORHEAD

North Pacific armorhead

Pseudopentaceros wheeleri

DESCRIPTION: Bluish brown, paler ventrally; reddish on head; orange on anal and pelvic fin spines; dark wavy lines or spots dorsolaterally in small individuals. Snout long and mouth small, maxilla not extending to below eye (1). Spines of dorsal and anal fins originating alternately on either side of the midline, and depressible in a groove; usually 4 anal fin spines (2); pectoral fin elongate, pointed. Small ctenoid (rough-edged) scales.
SIZE: Length to 53 cm TL.
DISTRIBUTION: Gulf of Alaska to central California. Far offshore waters; juveniles epipelagic, adults benthic; around seamount summits and pelagically in open ocean from surface to depth of 402 m.
(Mecklenburg et al. 2002)

Pseudopentaceros wheeleri

RONQUILS

Searcher

Bathymaster signatus

DESCRIPTION: Light olive brown to dark green dorsally, paler ventrally, with faint vertical bars; conspicuous black patch at anterior end of dorsal fin (1); bright orange pores on preopercle and behind maxilla (2). Dorsal fin long-based without a notch; tail rounded; lateral line straight; no anal fin spines; body slender.
SIZE: Length to 38 cm TL.
DISTRIBUTION: Bering Sea from Navarin Canyon and through Aleutian Islands to Washington. Bottom, on sand and gravel as well as near shore near rock and algae at depth of 25-300 m.
(Kessler 1985, Mecklenburg et al. 2002)

Alaskan ronquil

Bathymaster caeruleofasciatus

DESCRIPTION: Reddish brown, with or without blue vertical bars; head pores blackish and conspicuous; adults occasionally with dark blue or blackish spots on opercle; small specimens paler, often with dark blotch on opercle. Maxilla extends beyond eye (1). First dorsal rays unbranched; pectoral fin reaches origin of anal fin; caudal fin truncate. Scales ctenoid (rough-edged); caudal fin scaled for more than half its length; inner rays of pelvic fin scaled. Gill rakers 14 or fewer on lower limb of first arch.
SIZE: Length to 30 cm TL.
DISTRIBUTION: Aleutian Islands to British Columbia, and to Commander Islands. Subtidal, in rocky habitat, to depth of about 95 m.
(Mecklenburg et al. 2002)

OTHER RONQUILS. The ronquil family is composed of seven species; two others are found in Alaska waters. The smallmouth ronquil *Bathymaster leurolepis* is reddish-brown, its maxilla extends to below mid-eye, and it is found in southeastern Alaska to Pribilof Islands and along the Aleutian chain. The northern ronquil *Ronquilus jordani* male is orange dorsally and the female is olive green dorsally. It has scaly cheeks, the first 20-30 dorsal fin rays are not branched, and it often has thin yellow lines on its sides and yellow markings on fins and tail.

Bathymaster signatus

Bathymaster caeruleofasciatus

EELPOUTS

Shortfin eelpout

Lycodes brevipes

DESCRIPTION: Brown or gray-brown, paler ventrally; lower part of head cream or tan, in some specimens clearly demarcated from dark upper part; up to 13 narrow light bands on body and dorsal fin (1). Chin crests low, not united anteriorly. Head large, flat, with overhanging upper jaw. Tail pointed, not separated from dorsal and anal fins. Pelvic fin short, less than half of eye diameter (2); pectoral fin rounded or slightly indented, with upper rays longest.
SIZE: Length to 33 cm TL.
DISTRIBUTION: Western Bering Sea to Unalaska Island and to eastern Pacific off California. Sandy or muddy bottoms at depths of 27-973 m.
(Kessler 1985, Mecklenburg et al. 2002)

Wattled eelpout

Lycodes palearis

DESCRIPTION: Brown, lighter ventrally, usually with a black spot on the anterior end of the dorsal fin (1). Pelvic fins small; prominent cartilaginous ridges under head (2); pectoral fin rounded. Smaller specimens may be confused with *L. brevipes* as they both have narrow light bands on the body. *L. palearis* has light margins on the dorsal fin whereas *L. brevipes* has a dorsal fin with dark margins.
SIZE: Length to 54 cm TL.
DISTRIBUTION: Chukchi Sea, Bering Sea off the Aleutian Islands to Oregon. Muddy or sandy bottoms at depths of 25-925 m.
(Kessler 1985, Mecklenburg et al. 2002)

Lycodes brevipes

Lycodes palearis

Polar eelpout

Lycodes turneri

DESCRIPTION: Purple, with 10-12 bluish-white bands with purplish olive borders, or umber with cream bands and dark umber borders; underside of head and belly white; anterior bands broken or marbled by intrusion of ground color; first light band curves forward over head (1), second crosses anterior to dorsal fin and joins third band under pectoral fin; most bands remain distinct in adults. Head depressed. Pectoral fin rays 17 or 18.
SIZE: Length to 64 cm TL.
DISTRIBUTION: Beaufort and Chukchi seas to Bristol Bay in the eastern Bering Sea and to about Cape Olyutorskiy in the western Bering Sea. Soft bottoms, from 10-125 m.
(Mecklenburg et al. 2002)

Alaska eelpout

Bothrocara pusillum

DESCRIPTION: Flesh delicate and transparent, especially in head region. Eye diameter greater than snout length. Pelvic fin absent. Minute scales present on body, tail, and belly; scales absent from nape. Lateral line absent.
SIZE: Length to about 16 cm SL.
DISTRIBUTION: Eastern Bering Sea to Pacific Ocean off southern British Columbia. Bathydemersal; depth range 91-2,189 m.
(Mecklenburg et al. 2002)

OTHER EELPOUTS. There are six other eelpouts commonly observed in research surveys and as bycatch in commercial fisheries in Alaska, including *Lycodes polaris*, *L. raridens*, *L. diapterus*, *L. concolor*, *Bothrocara brunneum*, and *Lycodapus mandibularis*. Eelpouts that don't fit the above descriptions should be saved for positive identification.

Lycodes turneri

Bothrocara pusillum

PRICKLEBACKS

Decorated warbonnet

Chirolophis decoratus

DESCRIPTION: Shades of brown and orange with pale spots and bands; pelvic fin pale. It is distinguished by a single long cirrus above eyes that is more than twice as long as eye diameter (1); many cirri on top of head; few simple cirri on cheeks and sides of head. First 4-9 dorsal fin spines with multiple cirri (2); pelvic fin well developed.

SIZE: Length to 42 cm TL.

DISTRIBUTION: Eastern Bering Sea and Aleutian Islands to northern California. Rocky bottom and reef crevices, subtidal to depths of 91 m.

(Kessler 1985, Mecklenburg et al. 2002)

Longsnout prickleback

Lumpenella longirostris

DESCRIPTION: Dark bluish brown to olive brown; fin margins blackish. Snout elongate, projecting beyond and overhanging lower jaw (1); maxilla not extending to below eye (2). Dorsal fin spines usually stiff and with free tips (3); anal fin spines typically 3-5; tail rounded.

SIZE: Length to 41 cm TL.

DISTRIBUTION: Bering Sea and Aleutian Islands to southern British Columbia. Bottom, usually on outer shelf and upper slope, from depths of 25-1,140 m.

(Mecklenburg et al. 2002)

Snake prickleback

Lumpenus sagitta

DESCRIPTION: Light green to tannish or gray dorsally; cream ventrally. Irregular small dark blotches or streaks dorsolaterally with a line of dark, dash-like or oval marks along midbody (1). Brown bars and dots on caudal fin produce vertical bars. Body very elongate and slender. Maxilla extends to anterior edge of pupil. First few dorsal fin spines short and free; dorsal and anal fins not connected to caudal fin; almost always one anal fin spine.

SIZE: Length to 51 cm TL.

DISTRIBUTION: Bering Sea and Aleutian Islands to northern California. Sandy bottom with a mixture of silt, pebbles, and stones; from shore to depths of 425 m.

(Mecklenburg et al. 2002)

Chirolophis decoratus

Lumpenella longirostris

Lumpenus sagitta

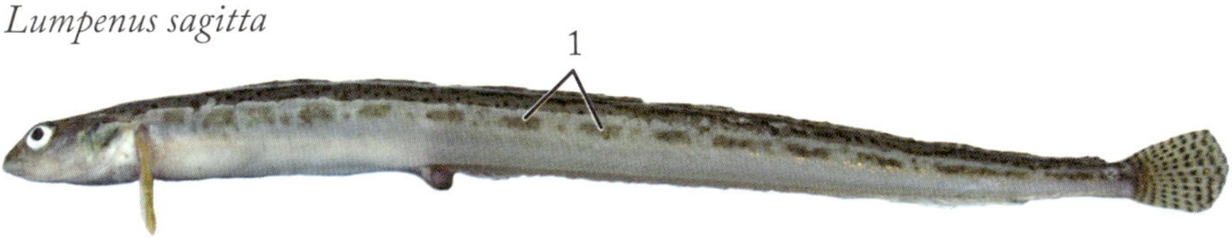

Whitebarred prickleback

Poroclinus rothrocki

DESCRIPTION: Light brown dorsally, pale ventrally; 10-12 narrow white vertical bars dorsolaterally on body (1); fins without distinct markings. Mouth small, maxilla extends to front of eye or pupil. Dorsal and anal fins not connected to caudal fin; anal fin with 3 spines; caudal fin rounded to nearly pointed.
SIZE: Length to 25 cm TL.
DISTRIBUTION: Southeastern Bering Sea and Aleutian Islands to southern California. Rocky bottom from 46-128 m.
(Mecklenburg et al. 2002)

Blackline prickleback

Acantholumpenus mackayi

DESCRIPTION: Body light yellowish to brownish; dark line on back at base of dorsal fin (1), 2 dark broken lines below; caudal fin dark, without bands. Body extremely elongate. Eye diameter less than snout length; maxilla extends to anterior edge of pupil or to mid-eye. Dorsal and anal fins not connected to caudal fin; anterior spines of dorsal fin short but not free.
SIZE: Length to 70 cm SL.
DISTRIBUTION: Southeastern Chukchi Sea, eastern Bering Sea, and Aleutian Islands; northeastern Gulf of Alaska. Mud or sand bottoms in shallow water near shore to depth of 56 m.
(Mecklenburg et al. 2002)

Stout eelblenny

Anisarchus medius

DESCRIPTION: Body creamy white or yellowish to reddish; marked with vague darker spots or monotone, sometimes with oblique red bars on dorsal fin; anal fin without markings; caudal fin banded. Dorsal and anal fins connected to caudal fin (1); caudal fin rounded; pelvic fin present. Gill membranes united far forward, attached to isthmus under anterior half of eye (2).
SIZE: Length to 18 cm TL.
DISTRIBUTION: Circumpolar, Beaufort Sea to Okhotsk and Japan seas and to southeastern Alaska. Almost exclusively on muddy bottoms from near shore to depth of 150 m.
(Mecklenburg et al. 2002)

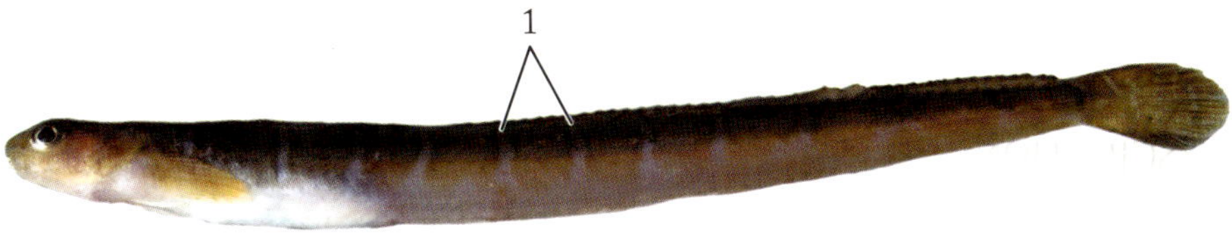

Poroclinus rothrocki

Acantholumpenus mackayi

Anisarchus medius

OTHER PRICKLEBACKS with blotches on side. The slender eelblenny *Lumpenus medius* has dorsal and anal fin rays that end at tail base; daubed shanny *L. maculatus* has irregular blotches that extend down lower side.

WRYMOUTHS

Giant wrymouth

Cryptacanthodes giganteus

DESCRIPTION: Brown or gray coloration. Eel-like body; upward-directed mouth (1); distinguished by its large size and the presence of scales.
SIZE: Length to 140 cm TL.
DISTRIBUTION: Southeastern Bering Sea, eastern Aleutian Islands, Gulf of Alaska to northern California. Soft bottom, often buried, at depths of 6-128 m.
(Kessler 1985, Mecklenburg et al. 2002)

Dwarf wrymouth

Cryptacanthodes aleutensis

DESCRIPTION: Bright red or pink from blood showing through skin; skin loose and transparent. Mouth strongly oblique, nearly vertical; scales absent.
SIZE: Length to 31 cm TL.
DISTRIBUTION: Southeastern Bering Sea and eastern Aleutian Islands to northern California. Soft bottoms, buried, at depths of 28-350 m.
(Mecklenburg et al. 2002)

1

Cryptacanthodes giganteus

Cryptacanthodes aleutensis

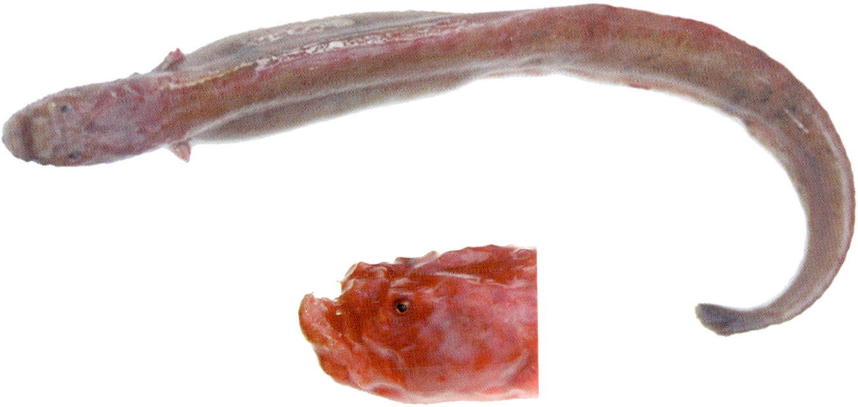

GUNNEL

Crescent gunnel

Pholis laeta

DESCRIPTION: Body yellowish green to orange-brown, paler ventrally; orange or reddish brown anal, caudal, and pectoral fins; pale streaks bordered with black or purple radiate from eye (1); brown or blackish, crescent-shaped to nearly circular markings enclosed in a yellow or orange area extending from upper surface of body onto dorsal fin (2); numerous pale blotches on side.
SIZE: Length to 25 cm TL.
DISTRIBUTION: Aleutian Islands east along Alaska Peninsula to Port Heiden, and Gulf of Alaska shores to northern California. Intertidal and shallow subtidal areas among eelgrass, algae, or rocks, to depth of 73 m.
(Mecklenburg et al. 2002)

WOLFFISH

Bering wolffish

Anarhichas orientalis

DESCRIPTION: Rosy brown, reddish brown, or black, mottled and blotched; juveniles with dark stripes on body. Body deep, eel-like; dorsal and anal fins not confluent with caudal fin; caudal fin gently rounded. Mouth large; prominent canine teeth (1).
SIZE: Length to 112 cm TL.
DISTRIBUTION: Beaufort and Chukchi seas to northern Gulf of Alaska. Coastal areas on gravel and sand bottom, not found deeper than about 100 m.
(Kessler 1985, Mecklenburg et al. 2002)

OTHER WOLFFISHES. The wolf-eel *Anarhichthys ocellatus* is distinguished by a long, slender body, covered with large dark spots surrounded by light color.

Pholis laeta

Anarhichas orientalis

Juvenile

PROWFISH

Prowfish

Zaprora silenus

DESCRIPTION: Adults are grayish blue to green, with or without darker spots and yellow blotches; young are orange-brown with dark and light spots and blotches. Large white pores (1); head high, blunt; body long, deep; tail straight; no pelvic fins. The dorsal and anal fins are especially high in the young.
SIZE: To 1 m SL or more.
DISTRIBUTION: Bering Sea to Pacific Ocean off San Miguel Island, California. Adults near bottom at depths to 675 m; young fish often taken near surface over deep waters.
(Kessler 1985, Mecklenburg et al. 2002)

SANDFISH

Pacific sandfish

Trichodon trichodon

DESCRIPTION: Brown dorsally, silver ventrally; dark brown on top of head extends back along base of dorsal fin as wide, short bars and along lateral line as unbroken or broken streak. Mouth strongly oblique to nearly vertical; maxilla extends to mid-eye. Caudal fin forked or strongly indented; pectoral fin broadly rounded when spread.
SIZE: Length to 31 cm TL.
DISTRIBUTION: Bering Sea and Aleutian Islands to central California. Intertidal to depth of 375 m.
(Mecklenburg et al. 2002)

Zaprora silenus

Juvenile

Trichodon trichodon

SAND LANCE

Pacific sand lance

Ammodytes hexapterus

DESCRIPTION: Metallic blue dorsally, silver ventrally. Long-based dorsal fin fits in a groove. Head pointed, with lower jaw protruding (1). Tail forked (2); no pelvic fins. Body long, slender.
SIZE: Length to 28 cm TL.
DISTRIBUTION: Western Canadian Arctic to Pacific Ocean off southern California. Intertidal to 100 m.
(Kessler 1985, Mecklenburg et al. 2002)

RIGHTEYE FLOUNDERS

Pacific halibut

Hippoglossus stenolepis

DESCRIPTION: Eyed side usually uniform dark gray or brown, sometimes with light blotches; blind side white. Mouth large; teeth prominent. Body slightly elongate; lateral line arched.
SIZE: Length to 267 cm TL.
DISTRIBUTION: Chukchi Sea to eastern Pacific Ocean, to northern Baja California. On a variety of bottom types at depths of 6-1,100 m.
(Kessler 1985, Mecklenburg et al. 2002)

Ammodytes hexapterus

Hippoglossus stenolepis

Greenland turbot

Reinhardtius hippoglossoides

DESCRIPTION: Eyed side dark purplish brown to black with a green sheen and small scales; blind side dark. Body elongate. Lower jaw large, projecting. Upper eye visible from blind side. Maxilla extends to below posterior part of eye or beyond eye. Preopercle with right angle at posterior margin and 4 or 5 pores near edge. Lateral line nearly straight. Gill rakers short and blunt.
SIZE: Length to 130 cm TL.
DISTRIBUTION: Beaufort and Chukchi seas to eastern Pacific Ocean off northern Baja California. Soft bottom at depths of 14-2,000 m.
(Mecklenburg et al. 2002)

Arrowtooth flounder

Atheresthes stomias

DESCRIPTION: Eyed side brown; scales darker at edges; blind side off-white to light gray. Upper eye visible from blind side. Body rather elongate. Maxilla extends to below posterior edge of eye or beyond. Teeth conical, sharp, widely spaced; teeth on upper jaw visible when mouth is closed. Gill rakers long and slender, more than 14 rakers on first arch; two rakers on top of the second arch. Lateral line slightly curved above pectoral fin (1). (Also known as *Reinhardtius stomias.*)
SIZE: Length to 86 cm TL.
DISTRIBUTION: Bering Sea to Aleutian Islands, and to Pacific Ocean off central California. Found on soft bottom at depths of 12-900 m.
(Mecklenburg et al. 2002)

Kamchatka flounder

Atheresthes evermanni

DESCRIPTION: Eyed side dark brown; blind side off-white. Upper eye not visible from blind side. Maxilla extends to below posterior edge of eye or beyond. Teeth conical, sharp, widely spaced, largest in front. Gill rakers long and slender; less than 14 gill rakers on first arch; one raker on top of second arch. Lateral line gradually rises to slightly curved above pectoral fin. (Also known as *Reinhardtius evermanni.*)
SIZE: Length to 110 cm TL.
DISTRIBUTION: Bering Sea to Aleutian Islands, and southwestern Gulf of Alaska. Found on soft bottom at depths of 25-1,200 m.
(Mecklenburg et al. 2002)

Reinhardtius hippoglossoides

Atheresthes stomias

Juvenile

Atheresthes evermanni

Blind side

Flathead sole

Hippoglossoides elassodon

DESCRIPTION: Eyed side brownish; dorsal and anal fins blotched; blind side translucent, with opaque white area in center; caudal fin membranes mostly clear. Body ovate; strongly compressed, especially in small fish. Lower jaw compressed, symphyseal knob present; mandibular articulation pointed; 15-24 gill rakers on lower first arch. Interorbital space narrow, with sharp ridge (1). Anal spine pronounced. Caudal fin rounded or broadly pointed. Lateral line slightly curved above pectoral fin or nearly straight.
SIZE: Length to 56 cm TL.
DISTRIBUTION: Bering Sea and Aleutian Islands to northern California. Silty or muddy bottoms from near shore to depth of about 1,050 m.
(Mecklenburg et al. 2002)

OTHER SPECIES OF *HIPPOGLOSSOIDES*. The Bering flounder *H. robustus* occupies the same range as *H. elassodon*. It differs from *H. elassodon* by not having blotched dorsal and anal fins, body is slightly more elongate, the space between eyes slightly wider and without a ridge, it has 9-13 gill rakers on lower first arch, and the caudal fin is round.

Butter sole

Isopsetta isolepis

DESCRIPTION: Eyed side brown, sometimes mottled or blotched; yellow on edges of dorsal and anal fins in freshly caught fish; blind side white; body ovate. Upper eye not visible from blind side. Maxilla extends to below anterior part of eye. Anal spine pronounced. Caudal fin rounded or broadly pointed. Lateral line with low arch above pectoral fin (1); accessory dorsal branch present, extending beyond operculum (2).
SIZE: Length to 55 cm TL.
DISTRIBUTION: Southern Bering Sea and Aleutian Islands to southern California. Muddy bottom at depths of 6-425 m.
(Mecklenburg et al. 2002)

Hippoglossoides elassodon

Juvenile

Isopsetta isolepis

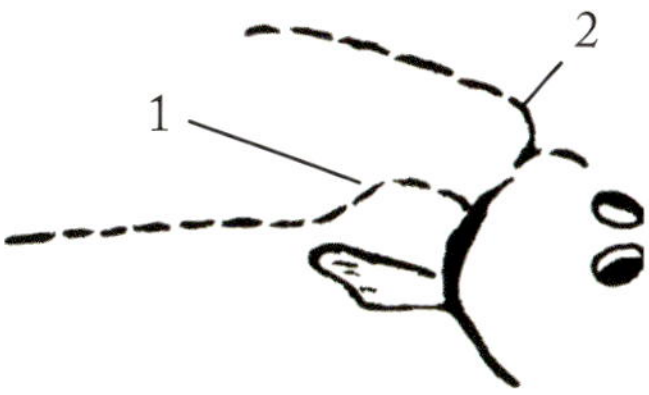

Rex sole

Glyptocephalus zachirus

DESCRIPTION: Eyed side light brown to gray; both sides with black speckles; pectoral fin mostly black; blind side off-white to dusky. Body elongate; limp, slimy; snout rounded. Maxilla extends to below anterior edge of eye or slightly beyond. Anal spine pronounced. Pectoral fin on eyed side longer than head; caudal fin rounded or double truncate. Lateral line nearly straight; no anterior branch.
SIZE: Length to 61 cm TL.
DISTRIBUTION: Western Bering Sea to southern tip of Kamchatka, Bering Sea and Aleutian Islands to central Baja California. Sandy or muddy bottom from shallow water near shore to depth of about 850 m.
(Mecklenburg et al. 2002)

Southern rock sole

Lepidopsetta bilineata

DESCRIPTION: Eyed side yellow to greenish-brown with scattered dark-brown blotches and light spots; brown blotches on median fins; blind side bright white, with muscle bands highlighted white (1). Interorbital space narrow (2); maxilla extends to below anterior portion of eye. Anal spine pronounced. Caudal fin rounded or double truncate. Lateral line strongly arched above pectoral fin (3); accessory dorsal branch extends below first 13-18 fin rays (see northern rock sole). Gill rakers broad and robust, total on first arch typically 10 or less.
SIZE: Length to 58 cm TL.
DISTRIBUTION: Southeastern Bering Sea and eastern Aleutian Islands to Baja California. Sand and gravel bottoms to depth of 339 m.
(Mecklenburg et al. 2002)

Glyptocephalus zachirus

Lepidopsetta bilineata

Northern rock sole

Lepidopsetta polyxystra

DESCRIPTION: Eyed side yellowish brown to greenish brown with scattered brown blotches and light spots. Blind side uniformly creamy white without glossy white highlights. Interorbital space wide (1). Lateral line strongly arched above pectoral fin (2) with an accessory dorsal branch (3); no distinct line between body and dorsal and anal fins. Scales on eyed side rough. Anal spine pronounced. Gill rakers slender and pointed, total on first arch typically 10 or more, on upper limb 3 or more.
SIZE: Length to 69 cm TL.
DISTRIBUTION: Bering Sea south and throughout Aleutian Islands to Washington. Sandy bottom to depth of 246 m.
(Mecklenburg et al. 2002)

Dover sole

Microstomus pacificus

DESCRIPTION: Eyed side brown, with indistinct darker mottling; fins blackish at edges; blind side smudgy off-white to dark brownish-gray. Body elongate, soft, and limp. Maxilla extends to below anterior part of eye. Lateral line nearly straight, with low curve above pectoral fin; short anterior branch. Median fins densely scaled. Teeth primarily on blind side; never more than 3 on eyed side.
SIZE: Length to 76 cm TL.
DISTRIBUTION: Southeastern Bering Sea and Aleutian Islands to southern Baja California. Muddy bottoms at depths of 9-1,244 m.
(Mecklenburg et al. 2002)

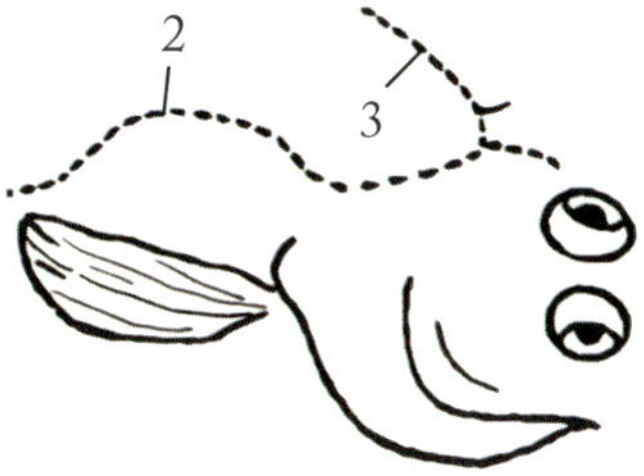

Lepidopsetta polyxystra

Microstomus pacificus

Yellowfin sole

Limanda aspera

DESCRIPTION: Eyed side dark brown or olive, fins often yellow; blind side white. Lateral line with a high flat-topped arch and no dorsal branch. Distinct dark line between body and dorsal and anal fins (1).
SIZE: Length to 49 cm TL.
DISTRIBUTION: Beaufort Sea to Atka Island in the Aleutian Islands, and Gulf of Alaska to British Columbia. Soft bottom at depths of 10-600 m.
(Kessler 1985, Mecklenburg et al. 2002)

English sole

Parophrys vetulus

DESCRIPTION: Eyed side uniform reddish brown or olive brown; vague blotching usually disappears in adults; blind side white to pale yellow. Body rather elongate; head and snout elongate and pointed. Portion of upper eye visible from blind side; high narrow ridge between eyes. Maxilla extends to below anterior edge of eye or a little beyond. Anal spine pronounced. Lateral line with low curve above pectoral fin; accessory dorsal branch long.
SIZE: Length to 61 cm TL.
DISTRIBUTION: Bering Sea to Alaska Peninsula and Aleutian Islands; Gulf of Alaska to central Baja California. Soft bottom from lower intertidal to depth of about 550 m.
(Mecklenburg et al. 2002)

Starry flounder

Platichthys stellatus

DESCRIPTION: Eyed side dark brown; blind side white; median fins white to orange with black bands. Body deep. Lateral line with slight curve above pectoral fin. Anal spine pronounced. Scattered rough tubercles on eyed side.
SIZE: Length to 91 cm TL.
DISTRIBUTION: Alaska Beaufort Sea to southern California. Soft bottoms from intertidal to depth of 375 m.
(Mecklenburg et al. 2002)

Limanda aspera

Parophrys vetulus

Platichthys stellatus

Alaska plaice

Pleuronectes quadrituberculatus

DESCRIPTION: Eyed side greenish gray to almost black; young adults spotted and blotched; blind side yellow. Body deep; postocular ridge with 4-7 bony knobs (1). Anal spine conspicuous; caudal fin rounded.
SIZE: Length to 62 cm TL.
DISTRIBUTION: Chukchi Sea through Bering Sea to Unalaska Island and Gulf of Alaska, to inside waters of southeastern Alaska. Soft bottoms at depths of 6-475 m.
(Mecklenburg et al. 2002)

Pleuronectes quadrituberculatus

SPONGES

The body of knowledge regarding sponges is increasing rapidly, so caution should be exercised with respect to identification, particularly when samples are unavailable for inspection. That said, identifications of many sponges are difficult to make from photographs, especially those where identifications need to be confirmed by spicule analyses. Questionable animals should be retained for positive identification.

GLASS SPONGE

Glass sponges are the most symmetrical and individualized of the sponges. In this group, some of the spicules are often fused to form a skeleton that may be lattice-like and built of long siliceous fibers.

Vase sponge

Aulosaccus schulzei

DESCRIPTION: Very large, barrel or vase-shaped sponge. Texture wool-like, rough, very thick, soft, external surface lacks "halo" of angel hair–like spicules; osculum (aperture) lacks large prominent spicules. Color white.
SIZE: Height to 80 cm.
DISTRIBUTION: Bering Sea and Gulf of Alaska. DEPTH: 200-2,000 m.
(Clark 2006)

Aulosaccus schulzei

CALCAREOUS SPONGE

Calcareous sponges are distinctive in having spicules composed of calcium carbonate.

Hairy vase sponge

Leucandra sp.

DESCRIPTION: Small tubular sponge. Texture soft and surface profusely covered with long, soft spicules; osculum large, open. Color white.
SIZE: Height to 5 cm.
DISTRIBUTION: Aleutian Islands. DEPTH: 60-200 m.
(Clark 2006)

VARIABLE SKELETON SPONGES

About 95% of sponge species are in this group, including most of the common sponges. The skeleton is variable.

Soccer ball sponge

Geodia mesotriaena

DESCRIPTION: Large solid sponge. Texture stone-like, surface rather smooth beneath the fuzzy, spiculose covering; upper central surface with several small oscula. Color white-gray or brown (due to sediment).
SIZE: Diameter to 30 cm.
DISTRIBUTION: Aleutian Islands to British Columbia. DEPTH: 10-200 m.
(Clark 2006)

Leucandra sp.

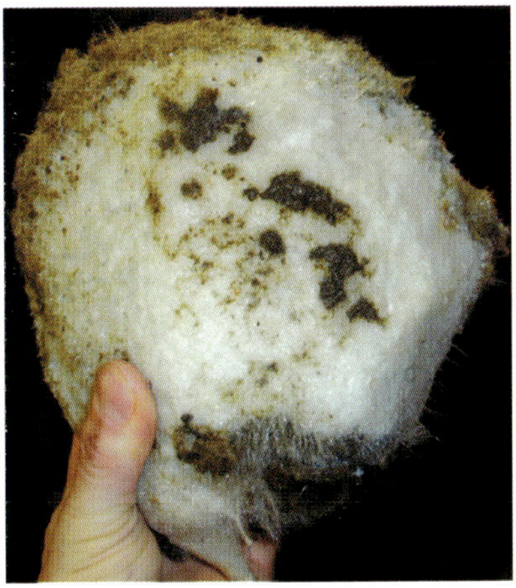

Geodia mesotriaena

Fuzzy cratered sponge
Vulcanella sp.

DESCRIPTION: Large, fuzzy appearing sponge with oscula in open smoothish craters or bite-like areas (1). Color dirty white, gray, or light brown.
SIZE: Diameter to 30 cm.
DISTRIBUTION: Aleutian Islands and Gulf of Alaska. DEPTH: 50-300 m.
(Clark 2006)

Furry ball sponge
Craniella spinosa

DESCRIPTION: Small fuzzy, ball-like sponge, covered with profuse, almost fur-like spicules. Color white.
SIZE: Diameter to 4 cm.
DISTRIBUTION: Aleutian Islands to British Columbia. DEPTH: 20-200 m.
(Clark 2006)

Baseball sponge
Craniella sp.

DESCRIPTION: Medium-sized, ball-like sponge. Surface rough, with protruding tips of spicules. Color white to gray. This sponge is often attached to other sponges, especially *Mycale* spp. and *Halichondria* spp.
SIZE: Diameter to 10 cm.
DISTRIBUTION: Aleutian Islands to Gulf of Alaska. DEPTH: 50-300 m.
(Clark 2006)

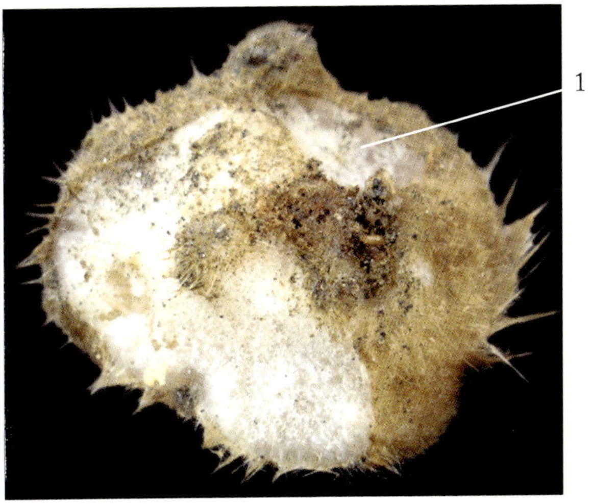

Vulcanella sp.

Craniella spinosa

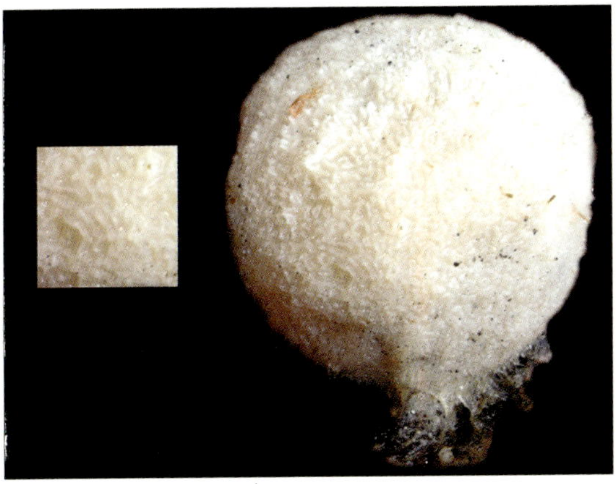

Craniella sp.

Hermit sponges

Suberites ficus, Suberites domuncula

DESCRIPTION: *Suberites ficus* is a firm, oval-shaped colony, with a smooth rubbery texture, which grows over a snail shell that eventually dissolves. This sponge is usually associated with the whiteknee hermit crab. Color is brown, yellow, or brick-red. *Suberites domuncula* is a small irregular sponge, often lumpy rather than smooth-textured. Color yellowish to red.
SIZE: Length to 15 cm.
DISTRIBUTION: *S. ficus* common north and south of the Alaska Peninsula. *S. domuncula* found from Bering Sea to British Columbia. DEPTH: 20-200 m.
(Kessler 1985, Clark 2006)

Nipple sponge

Polymastia sp.

DESCRIPTION: Large, encrusting sponge. Colonies with a broad, irregular flattened base bearing numerous short, stout oscular nipples. Color light yellow to orange.
SIZE: Width to 30 cm.
DISTRIBUTION: Aleutian Islands to Gulf of Alaska. DEPTH: 0-200 m.
(Clark 2006)

Tree sponge

Mycale loveni

DESCRIPTION: Very large, lobed sponge. Colony is massive, soft, thick, and erect. The skeleton is fibrous, and tree-like. Color yellow to tan.
SIZE: Height to 1 m.
DISTRIBUTION: Circumboreal, south to Oregon. DEPTH: 30-400.
(Clark 2006)

Suberites ficus

Suberites domuncula

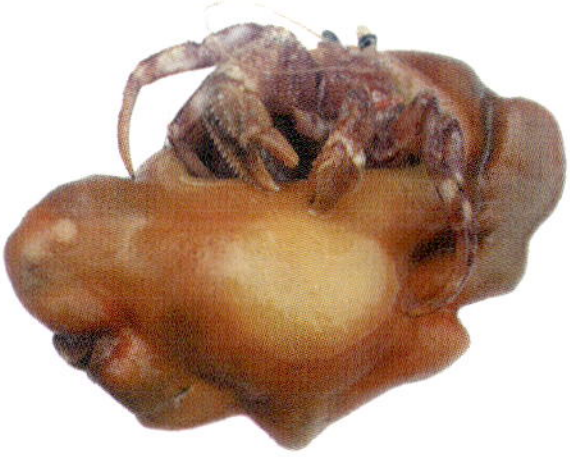

Polymastia sp.

Mycale loveni

Cotton ball sponge
Mycale lobata

DESCRIPTION: Small spherical sponge (1). The colonies are velvety to the touch, resembling cotton balls. The colony lives attached to branching hydroid colonies (2). Color tan.
SIZE: Diameter to 4 cm.
DISTRIBUTION: Aleutian Islands to Washington. DEPTH: 50-200 m.
(Clark 2006)

Lattice sponge
Melonchela clathriata

DESCRIPTION: Medium-sized, fan-shaped sponge. The erect colony is rarely branching and is perforated with diamond-like holes. Color tan.
SIZE: Height to 15 cm.
DISTRIBUTION: North Pacific Ocean and Bering Sea. DEPTH: 350-500+ m.
(Clark 2006)

Spud sponge
Histodermella sp.

DESCRIPTION: Medium-sized, potato-like sponge. Colonies are spherical, very firm; surface smooth with several projecting tubes. Color brown.
SIZE: Diameter to 10 cm.
DISTRIBUTION: Aleutian Islands to Gulf of Alaska. DEPTH: 100-300 m.
(Clark 2006)

Mycale lobata

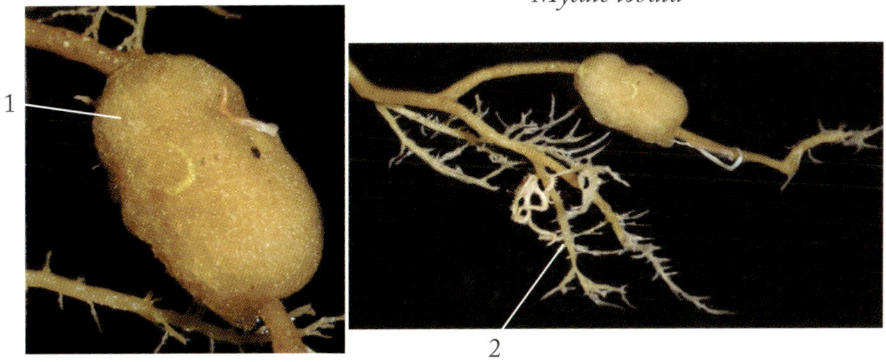

Melonchela clathriata

Histodermella sp.

Firm fingered sponge

Neoesperiopsis sp.

DESCRIPTION: Large branching sponge. Colonies are erect, with long tube-like branching; texture rather smooth, but bearing numerous small oscula along side of tubes. Color green, yellow, brown, or reddish.
SIZE: Height to 45 cm.
DISTRIBUTION: Bering Sea to British Columbia. DEPTH: 10-200 m.
(Clark 2006)

Rough China hat sponge

Neoesperiopsis infundibula

DESCRIPTION: Large, bowl-like sponge. Colonies are soft, funnel-shaped; texture fine-celled, concentric outgrowths on the exterior surface. Color yellow-brown.
SIZE: Diameter to 50 cm.
DISTRIBUTION: Bering Sea to British Columbia. DEPTH: 50-300 m.
(Clark 2006)

Soft brown sponge

Myxilla sp.

DESCRIPTION: Medium-sized, soft, globular sponge; texture very rough. Color brownish.
SIZE: Diameter to 30 cm.
DISTRIBUTION: Aleutian Islands to British Columbia. DEPTH: 20-200 m.
(Clark 2006)

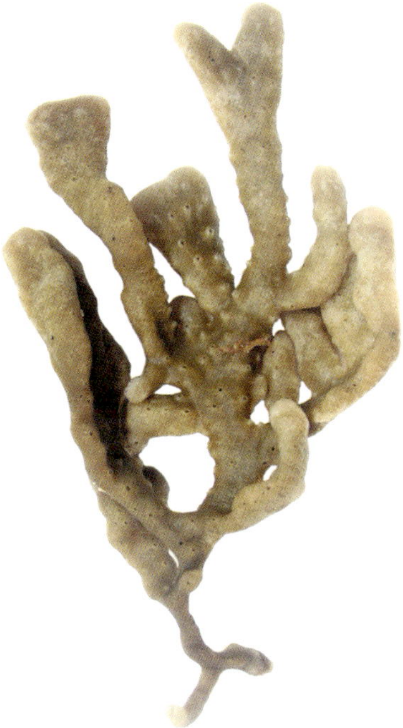

Neoesperiopsis sp.

Neoesperiopsis infundibula

Myxilla sp.

Scallop sponge
Myxilla incrustans

DESCRIPTION: Usually found growing on scallop shells of the genus *Chlamys*, especially the reddish scallop. Color bright yellow to brown.
SIZE: To 2 cm thick.
DISTRIBUTION: Arctic Ocean to California living symbiotically on scallop shells.
DEPTH: 10-200 m.
(Kessler 1985, Clark 2006)

HYDROIDS

Champagne flute hydroid
Bonneviella sp.

DESCRIPTION: Small clump-like colonies bearing large, slender, stalked, cup-like hydrothecae (1). Color light orange or pinkish.
SIZE: Hydrotheca length to 2 cm.
DISTRIBUTION: Bering Sea to Gulf of Alaska. DEPTH: 20-100 m.
(Clark 2006)

Ostrich-plume hydroid
Aglaophenia sp.

DESCRIPTION: Large, feather-like colonies. Branches are long, slender, bearing polyps on short, feather-like side branches. Color orange, brown, or tan.
SIZE: Height to 15 cm.
DISTRIBUTION: Bering Sea to Mexico. DEPTH: 0-160 m.
(Kozloff 1987, Clark 2006)

Myxilla incrustans

Bonneviella sp.

Aglaophenia sp.

HYDROCORALS

Stylaster hydrocoral

Stylaster campylecus parageus

DESCRIPTION: Medium-sized coral, with upright branching, rather prickly colonies. Main branches large, thick, branchlets irregular and twig-like, growing in several planes; cyclosystems small, protruding. Color orange-pink to nearly white.
SIZE: Height to 35 cm.
DISTRIBUTION: Aleutian Islands to British Columbia. DEPTH: 30-300 m.
(Wing and Barnard 2004, Clark 2006)

Common hydrocoral

Stylaster campylecus campylecus

DESCRIPTION: Medium-sized upright coral, with fan-like colonies; branching on one plane. Main branches large, thick, somewhat flattened, smaller branches rather robust, cyclosystems protruding, scarce on back of colony, numerous on front and edges. Color buff to pink or pink-orange.
SIZE: Height to 35 cm.
DISTRIBUTION: Aleutian Islands to Southeastern Alaska. DEPTH: 30-300 m.
(Clark 2006)

Pink hydrocoral

Stylaster verrillii

DESCRIPTION: Small upright coral, with blunt, thick-branched colonies. Cyclosystems not protruding, evenly spaced on all surfaces. Color pink to violet.
SIZE: Height to 5 cm.
DISTRIBUTION: Aleutian Islands to Washington. DEPTH: 87-366 m.
(Wing and Barnard 2004, Clark 2006)

raised cyclosystems

Stylaster campylecus parageus

Stylaster campylecus campylecus

Stylaster verrillii

Capped hydrocoral

Crypthelia trophostega

DESCRIPTION: Medium-sized upright, fan-like colony. Branches usually robust; cyclosystems large, evenly spaced, much more numerous on front of colony than on back, and bearing cap-like fixed lids (1). Color white or pale buff.
SIZE: Height to 20 cm.
DISTRIBUTION: Aleutian Islands. DEPTH: 300-800+ m.
(Wing and Barnard 2004, Clark 2006)

Rosy-undulate hydrocoral

Cyclohelia lamellata

DESCRIPTION: Colony relatively thin, lamellar, ruffled, or rose-shaped; gastropore and dactylopores rather evenly distributed on all surfaces. Color reddish or orange-pink, whitish at edges.
SIZE: Height to 10 cm.
DISTRIBUTION: Aleutian Islands. DEPTH: 100-500+ m.
(Wing and Barnard 2004, Clark 2006)

Rough hydrocoral

Errinopora zarhyncha

DESCRIPTION: Medium-sized, upright, branching colonies. Branches very robust, very rough on all surfaces; cyclosystems crowded. Dactylopore scoops very profuse and long (1-2 mm), and crowded on all surfaces. Color orange to pale orange.
SIZE: Height to 30 cm.
DISTRIBUTION: Aleutian Islands. DEPTH: 400-650+ m.
(Wing and Barnard 2004, Clark 2006)

Crypthelia trophostega

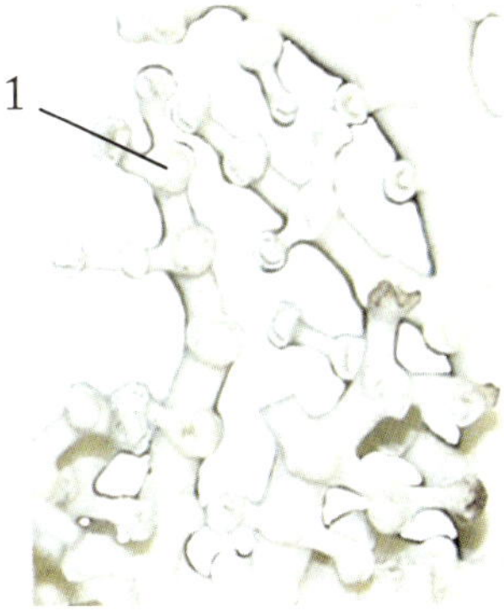

Cyclohelia lamellata

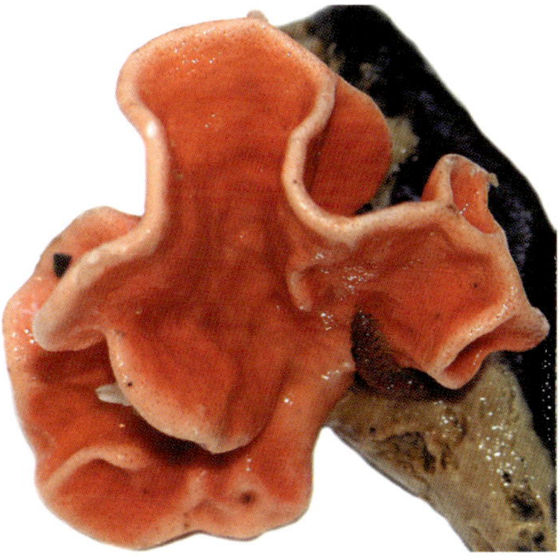

Errinopora zarhyncha

OCTOCORALS (SOFT CORALS)

Red mushroom coral

Anthomastus sp.

DESCRIPTION: Rather small mushroom-shaped colonies with a thick stalk and broad, hemispherical cap. Cap bears large retractable polyps (1). Color red, pink, yellow, or tan; polyps may be lighter or darker than cap.
SIZE: Height to 10 cm.
DISTRIBUTION: Aleutian Islands to California. DEPTH: 20-2,000 m.
(Wing and Barnard 2004, Clark 2006)

Gray mushroom coral

Anthomastus sp.

DESCRIPTION: Small mushroom-like coral. Stalk is short, stout, cap relatively large, hemispherical, bearing several to many large polyps. Color gray, sometimes pale yellow-orange polyps.
SIZE: Height to 10 cm.
DISTRIBUTION: Aleutian Islands to Gulf of Alaska. DEPTH: 100-1,000+ m.
(Clark 2006)

Sea raspberry

Gersemia sp.

DESCRIPTION: Small, lumpy coral. Colonies brain-like shaped when contracted; no hard skeleton. Color white to red.
SIZE: Contracted diameter to 10 cm.
DISTRIBUTION: Circumboreal, south to California. DEPTH: 10-200 m.
(Clark 2006)

Anthomastus sp.

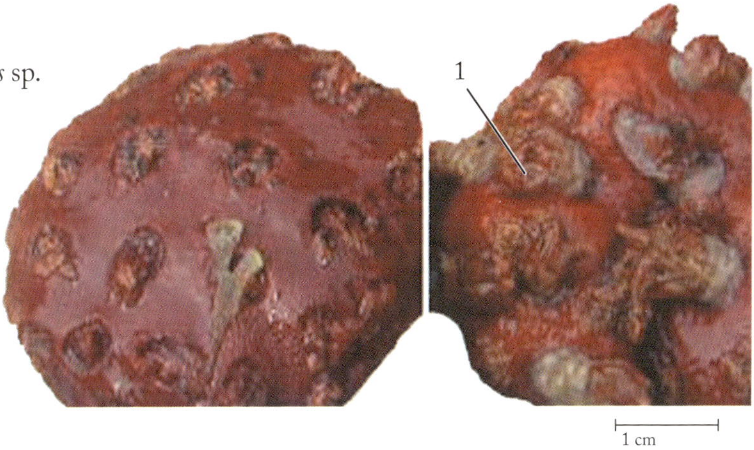

Anthomastus sp.

Gersemia sp.

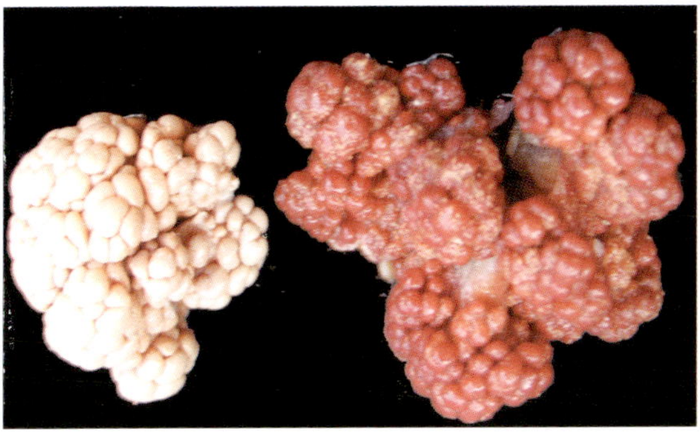

Contracted

Expanded

Purple soft coral

Anthothela grandiflora

DESCRIPTION: Very small encrusting or short-stalked coral; polyps relatively large. Color purple to violet.
SIZE: Height to 2 cm.
DISTRIBUTION: Circumboreal, south to Aleutian Islands. DEPTH: 50-150 m.
(Clark 2006)

Bubblegum coral

Paragorgia arborea

DESCRIPTION: Very large coral. Colonies tree-like, branching in several planes. Polyps small, rosette-shaped, well-spaced, evertible. Branches are massive, rounded; skeleton soft, easily broken. Color brick red, orange, yellow, or cream.
SIZE: Height to 3 m.
DISTRIBUTION: Circumboreal, south to California. DEPTH: 30-1,000+ m.
(Wing and Barnard 2004, Clark 2006)

Sea fan

Calcigorgia sp.

DESCRIPTION: Medium-sized fan-shaped coral. Branches irregular, flexible, covered with numerous knobs from which the contractile polyps emerge. Color pink, pale orange, pink-orange or white.
SIZE: Height to 30 cm.
DISTRIBUTION: Aleutian Islands. DEPTH: 20-100+ m.
(Wing and Barnard 2004, Clark 2006)

Anthothela grandiflora

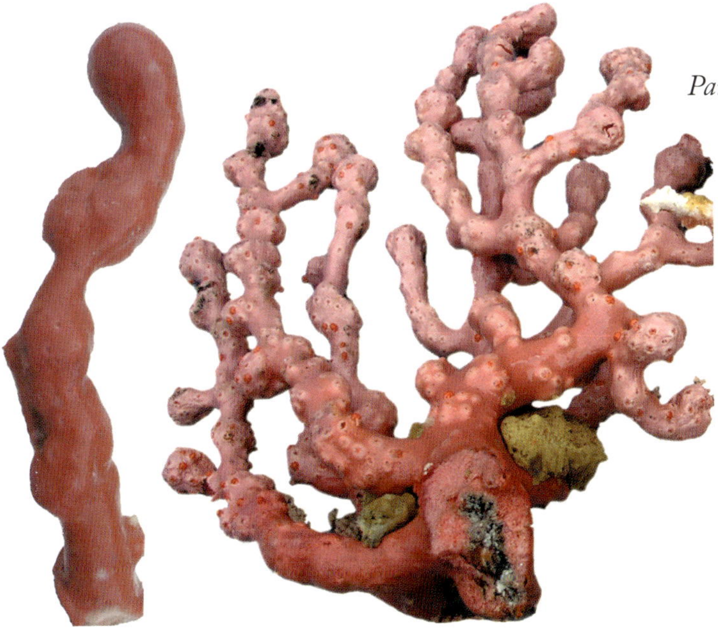

Paragorgia arborea

Calcigorgia sp.

Purple sea fan

Muriceides nigra

DESCRIPTION: Medium-sized fan or tree-shaped coral. Branches irregular, thin, flexible, profusely covered with non-contractile polyps. Color olive-brown to blackish with purple polyps.
SIZE: Height to 30 cm.
DISTRIBUTION: Aleutian Islands. DEPTH: 50-400 m.
(Wing and Barnard 2004, Clark 2006)

Kool's candelabrum gorgonian

Cryogorgia koolsae

DESCRIPTION: Medium-sized candelabrum-shaped coral. Branching usually only on one plane; branches relatively long, thick, blunt; polyps partially contractile. Color pale orange to pink-orange.
SIZE: Height to 40 cm.
DISTRIBUTION: Aleutian Islands. DEPTH: 50-150 m.
(Clark 2006)

Pacific red sea fan

Swiftia sp.

DESCRIPTION: Small tree-like gorgonian. Branches flexible, rubbery-appearing, not anastomozing (interconnecting); polyps arising from small mounds. Color red.
SIZE: Height to 15 cm.
DISTRIBUTION: Aleutian Islands to California. DEPTH: 18-300 m.
(Clark 2006)

Muriceides nigra

Cryogorgia koolsae

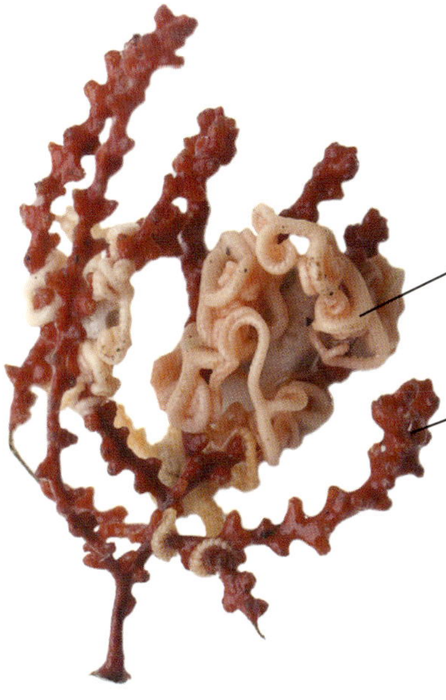

Pinnate amphilaphis

Amphilaphis sp.

DESCRIPTION: Large feather-like gorgonian. Branches long and slender, with pinnate branchlets; polyps crowded, rather slender. Color orange to orange-red.
SIZE: Height to 75 cm.
DISTRIBUTION: Aleutian Islands. DEPTH: 50-150 m.
(Clark 2006)

Plumarella coral

Plumarella sp.

DESCRIPTION: Small plumose gorgonian; fan-like colony. Branches slender; marginal scales, short spines near apexes. Live polyps white, pale orange to pink-orange.
SIZE: Height to 25 cm.
DISTRIBUTION: Aleutian Islands. DEPTH: 100-600 m.
(Wing and Barnard 2004, Clark 2006)

Pacific red tree coral

Primnoa pacifica

DESCRIPTION: Very large gorgonian. Branches profusely covered with large (up to 6.5 mm) polyps; polyps not greatly expanded distally. Color pink.
SIZE: Height to 2.5 m.
DISTRIBUTION: Japan to California. DEPTH: 9-300 m.
(Wing and Barnard 2004, Clark 2006)

Amphilaphis sp.

Plumarella sp.

Primnoa pacifica

Utinomi's armored sea fan

Arthrogorgia utinomii

DESCRIPTION: Medium-sized gorgonian. Tree-like colony; branches in a single plane, branches sparse, robust; polyps large, robust, close-set, box-like; basal scales of polyps bearing long, broad projecting spines. Color pale orange to pinkish.
SIZE: Height to 35 cm.
DISTRIBUTION: Aleutian Islands. DEPTH: 400-800+ m.
(Wing and Barnard 2004, Clark 2006)

Superb thouarella

Thouarella superba

DESCRIPTION: Medium-sized coarse gorgonian. Branches long, relatively robust, closely packed and fan-like; branchlets short, rather stiff, especially on fronts and backs of colonies. Polyps are small, profuse. Color light orange to pink-orange.
SIZE: Height to 60 cm.
DISTRIBUTION: Aleutian Islands. DEPTH: 250-500+ m.
(Wing and Barnard 2004, Clark 2006)

Giant fanellia

Fanellia compressa

DESCRIPTION: Large pinnate gorgonian. Candelabra-like colony; branches long, slender, branchlets long; polyps arranged in closely placed whorls, six to eight polyps per whorl distally, about 6 to 7 whorls per cm. Color light orange to pink orange.
SIZE: Height to 1 m.
DISTRIBUTION: Eastern Aleutian Islands. DEPTH: 100-900 m.
(Wing and Barnard 2004, Clark 2006)

Arthrogorgia utinomii

Thouarella superba

Fanellia compressa

CUP CORAL

Alaska cup coral

Caryophyllia alaskensis

DESCRIPTION: Chalice small, relatively robust, with slender to robust pedicel; cone-shaped; septa numerous (1), center of chalice with large columella (column-like structure). Animal pale orange, pink, or red.
SIZE: Diameter to 2.5 cm.
DISTRIBUTION: Kamchatka to British Columbia. DEPTH: 100-500+ m.
(Wing and Barnard 2004, Clark 2006)

OTHER CUP CORALS. There are seven additional species of cup corals in the region. They differ in size of chalice, pedicel, and columella; coloration; and number of septa. For a key and more information see references above.

SEA PEN

Orange sea pen

Ptilosarcus gurneyi

DESCRIPTION: Thick sea pen, large when expanded; body very stout. Leaves are large, broad, and fleshy in two lateral rows. Color orange.
SIZE: Height to 60 cm.
DISTRIBUTION: Bering Sea to Mexico. DEPTH: 10-200 m.
(Kozloff 1987, Clark 2006)

SEA WHIP

Giant sea whip

Halipteris willemoesi

DESCRIPTION: Very large and thick. Base of colony forms a peduncle (1) that anchors it to the seafloor. Exposed portion of colony up to 1.5 m in height and supports fleshy secondary polyps (2) that cover the hard shaft, or rachis. Rachis is very long, leaves short, crowded; polyps in densely spiculose chalices. Color flesh to whitish.
SIZE: Height to 2.5 m.
DISTRIBUTION: Bering Sea to Gulf of Alaska. DEPTH: 30-300+ m.
(Clark 2006)

Caryophyllia alaskensis

Ptilosarcus gurneyi

Halipteris willemoesi

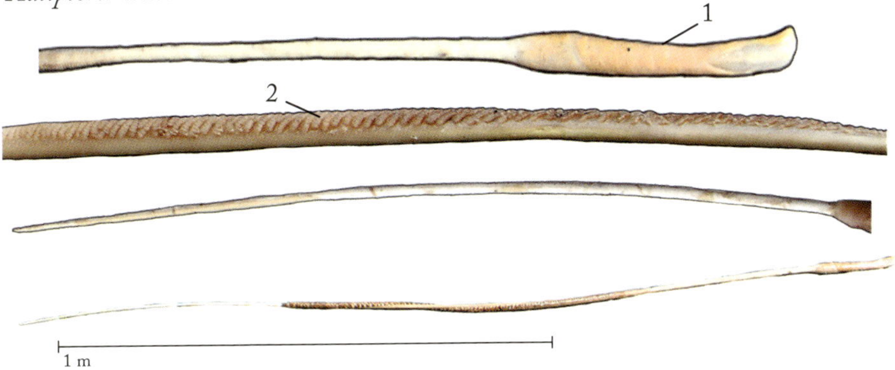

SEA ANEMONES

Mottled anemone

Urticina crassicornis

DESCRIPTION: The slender tentacles have a blunt, bulbous tip and are white with a red band (1). There are also red bands on the oral disc surrounding each tentacle. The column is usually a mottled red and olive, but can sometimes be mottled red, green, yellow, or solid colored. (Also known as *Urticina felina.*)
SIZE: Tentacular crown to about 20 cm.
DISTRIBUTION: Alaska to California. DEPTH: 0-50 m.
(Gotshall 1994)

Chevron-tentacled anemone

Cribrinopsis fernaldi

DESCRIPTION: Large anemone; column tall with many small white dots, relatively slender; tentacles long and covered with chevron-like sculpturing. Prevailing color of column, disk, and tentacles white.
SIZE: Diameter to 20 cm.
DISTRIBUTION: Bering Sea to Washington. DEPTH: 1-300 m.
(Kozloff 1987, Clark 2006)

Urticina crassicornis

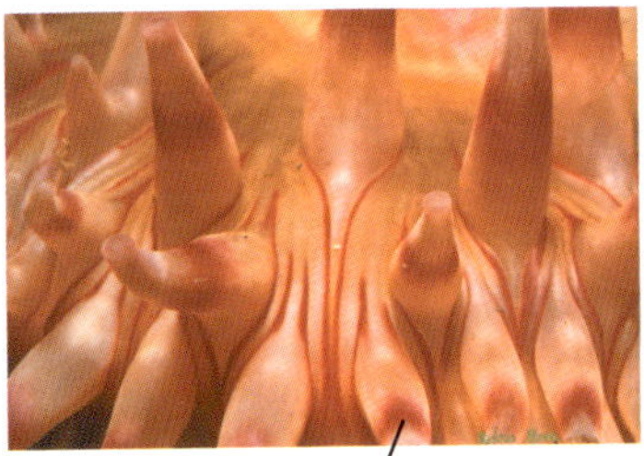

1

Cribrinopsis fernaldi

Gigantic (white-plumed) anemone

Metridium farcimen

DESCRIPTION: Large slimy anemone; column smooth; tentacles small and appear plumose (feathery) when expanded, cauliflower-like when contracted. Color cream, tan, brown, orange, or green.
SIZE: Diameter to 15 cm.
DISTRIBUTION: Bering Sea to California. DEPTH: 5-500 m.
(Gotshall 1994, Clark 2006)

JELLYFISH

Sunrise jellyfish

Chrysaora melanaster

DESCRIPTION: Adult is gelatinous consisting of a yellow-brown, bell-shaped mass with 16 evenly spaced, brownish rays radiating to the margin.
SIZE: Diameter to 60 cm.
DISTRIBUTION: Bering Sea, Aleutian Islands, and Gulf of Alaska.
DEPTH: Common in near-surface waters.
(Kozloff 1987, Clark 2006)

Metridium farcimen

Contracted

Chrysaora melanaster

SEGMENTED WORMS

Sea mouse

Aphrodita negligens

DESCRIPTION: Large, broad polychaete with a felt-textured surface and stiff bristle-like setae (1). Rolls into a clod-like ball when disturbed (2). Color gray to gray-brown.
SIZE: Length to 10 cm.
DISTRIBUTION: Bering Sea to Mexico. DEPTH: 20-165 m.
(Kessler 1985, Clark 2006)

Depressed scale worm

Eunoe depressa

DESCRIPTION: Small scale worm; scale tubercles scattered uniformly over scale surface. Often found on sponges. Color white or tan with light brown mottling.
SIZE: Length to 5 cm.
DISTRIBUTION: Bering Sea to Gulf of Alaska. DEPTH: 10-100+ m.
(Kessler 1985, Clark 2006)

Giant scale worm

Eunoe nodosa

DESCRIPTION: Large scale worm; scale tubercles in narrow band near outer scale margin (1). Often found in empty snail shells. Greenish brown color.
SIZE: Length to 10 cm.
DISTRIBUTION: Bering Sea to Gulf of Alaska. DEPTH: 10-100+ m.
(Kessler 1985, Clark 2006)

Aphrodita negligens

Eunoe depressa

Eunoe nodosa

Red-banded commensal worm

Arctonoe vittata

DESCRIPTION: Series of scales along the edges of the back. Some individuals are free living, and some live commensally. Hosts include the rough keyhole limpet, giant Pacific chiton, and 9 species of sea stars. Color is white or pale yellow, with a single broad transverse reddish-brown band.
SIZE: Length to 10 cm.
DISTRIBUTION: Bering Sea to Ecuador. DEPTH: Mid-intertidal to 275 m.
(O'Clair and O'Clair 1998, Clark 2006)

Northern feather duster worm

Eudistylia vancouveri

DESCRIPTION: Worm encased in a tough, horny tube (1). Tentacular crown dark maroon with dark green stripes (2).
SIZE: Tube length to 45 cm, 1 cm wide; head (tentacles) expands to 5 cm in diameter.
DISTRIBUTION: Alaska to southern California. DEPTH: Shallow subtidal, on rocks and dock pilings.
(Gotshall 1994)

Northern iridescent worm

Hemipodus borealis

DESCRIPTION: Long, segmented, cylindrical body tapers to a sharp point at each end. Club-shaped proboscis everted readily (1). Parapodia are elongate in front of the fans of setae. Color pinkish or purplish and iridescent.
SIZE: Length to 10 cm.
DISTRIBUTION: Alaska to Mexico. DEPTH: Mid to low intertidal and subtidal to 46 m.
(O'Clair and O'Clair 1998)

Eudistylia vancouveri

Hemipodus borealis

Calcareous tube worm

Serpula vermicularis

DESCRIPTION: Small brightly colored polychaete inside a secreted eccentrically coiled calcareous tube (1). Crown of feather-like plumes protrude from tubes when animal is active. When disturbed, the worm retreats into its tubes for protection. Well-developed, funnel-shaped operculum. Worm yellow with reddish plumes and white tubes.
SIZE: Length to 6 cm.
DISTRIBUTION: Bering Sea to California. DEPTH: 0-100 m.
(Kozloff 1987, Clark 2006)

OTHER WORMS. The worms shown in this book are more commonly encountered than other worms found in Alaska survey areas. Other worms occur, and many of them require microscopic techniques for positive identification.

LEECH

Striped sea leech

Notostomum cyclostomum

DESCRIPTION: Medium-sized smooth, tubular worm with large sucker disk at anterior end (1). Color purple-gray to brown, banded with cream. Found on king or Tanner crabs, where they often lay their eggs (2).
SIZE: Length to 10 cm.
DISTRIBUTION: Arctic Ocean to Gulf of Alaska. DEPTH: 10-370 m.
(Kessler 1985, Clark 2006)

CHITONS

Black katy

Katharina tunicata

DESCRIPTION: Body elongate, oval. Girdle black, thick, shiny and leathery, covering all but middorsal area. Valves embedded in girdle (1). Valves are white, but often discolored a dingy brownish or have various encrusting algae or animals growing on them; ventral surface is orange.
SIZE: Length to 13 cm.
DISTRIBUTION: Aleutian Islands, Kodiak Island, and southeastern Alaska to southern California. DEPTH: Intertidal and shallow subtidal. Usually attached to rocks in areas of heavy wave action.
(Barr and Barr 1983, O'Clair and O'Clair 1998)

Serpula vermicularis

Notostomum cyclostomum

Katharina tunicata

Giant Pacific (gumboot) chiton

Cryptochiton stelleri

DESCRIPTION: Valves white, butterfly shaped and completely covered by a thick, tough, red-brown girdle; surface of girdle roughened by minute spines. Usually found on rock, it is the largest chiton in the world.
SIZE: Length to 35 cm
DISTRIBUTION: Aleutian Islands to southeastern Alaska, south to California.
DEPTH: Low intertidal to 60 m.
(Barr and Barr 1983, Clark 2006)

White veiled chiton

Placiphorella pacifica

DESCRIPTION: Small, oval chiton; valves much wider than long. Girdle greatly expanded anteriorly and bearing sparse, short setae, often only at periphery (1). Color of valves and girdle whitish.
SIZE: Length to 4 cm.
DISTRIBUTION: Bering Sea to Chile. DEPTH: 80-2,000+ m.
(Clark 2006)

Bering chiton

Lepidozona beringiana

DESCRIPTION: Medium-sized elongate-to-oval chiton with faintly sculptured valves; central area of valves with fine, faint net-like reticulation. Radial areas smooth or with a few rows of fine granules. Girdle covered with fine, pointed scales. Color brick-red, tan, and white, often with reddish speckles.
SIZE: Length 3-4 cm.
DISTRIBUTION: Aleutian Islands. DEPTH: 70-400 m.
(Clark 2006)

OTHER CHITONS. There are more than 100 intertidal and subtidal chitons found along the Pacific coast from the Aleutian Islands south to Baja California. Chitons are slow moving and creep over rocks. Able to withstand the surf in exposed locations, they are also common in more protected waters. Chitons are not commonly found in pots or trawls unless attached to rocks or other organisms. For information on other chitons see Harbo 1997 and O'Clair and O'Clair 1998.

Cryptochiton stelleri

Placiphorella pacifica

Lepidozona beringiana

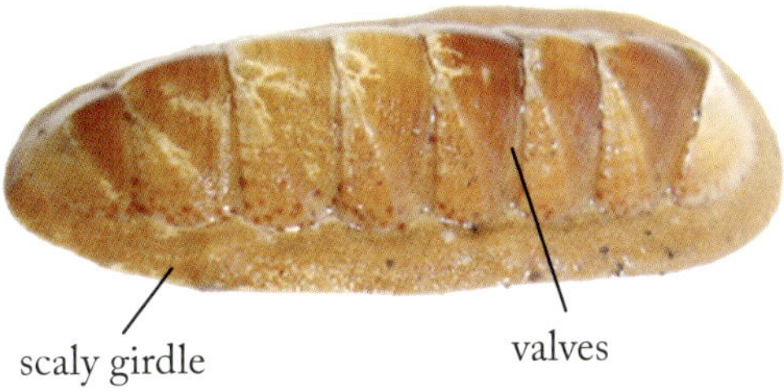

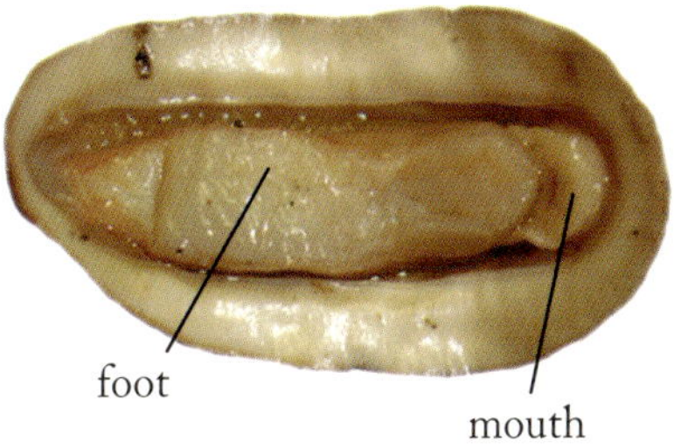

LIMPETS

Rough keyhole limpet

Diodora aspera

DESCRIPTION: Shell base oval, apex moderately well-elevated, with a small oval or nearly circular hole in summit (1). Surface ornamented by numerous low radial driblets cut by growth lines. Margins finely crenulate. Color white or gray, often with purplish-brown rays; interior white.
SIZE: Length to 6 cm.
DISTRIBUTION: Gulf of Alaska to Baja California. DEPTH: 0-40 m.
(Morris 1966, Clark 2006)

Whitecap limpet

Acmaea mitra

DESCRIPTION: Conical, cap-like shell with a circular base. Apex bluntly pointed and nearly central in position. Color dull white inside and out, some shells tinged with pink or pale green.
SIZE: Length to 3 cm.
DISTRIBUTION: Bering Sea to Baja California. DEPTH: Low intertidal zone areas with moderate to heavy wave action, to a subtidal depth of 34 m.
(Morris 1966)

SLIPPERSNAIL

Great slippersnail

Crepidula grandis

DESCRIPTION: Shell flat with large aperture (1). Shell body ends in a small spire with only one revolution (2). Often found attached to empty shells.
SIZE: Height to 5 cm.
DISTRIBUTION: Bering Sea and Aleutian Islands. DEPTH: 0-100 m.
(Kessler 1985, Clark 2006)

Diodora aspera

Acmaea mitra

Crepidula grandis

EXTERNAL FEATURES OF SOME GASTROPODS

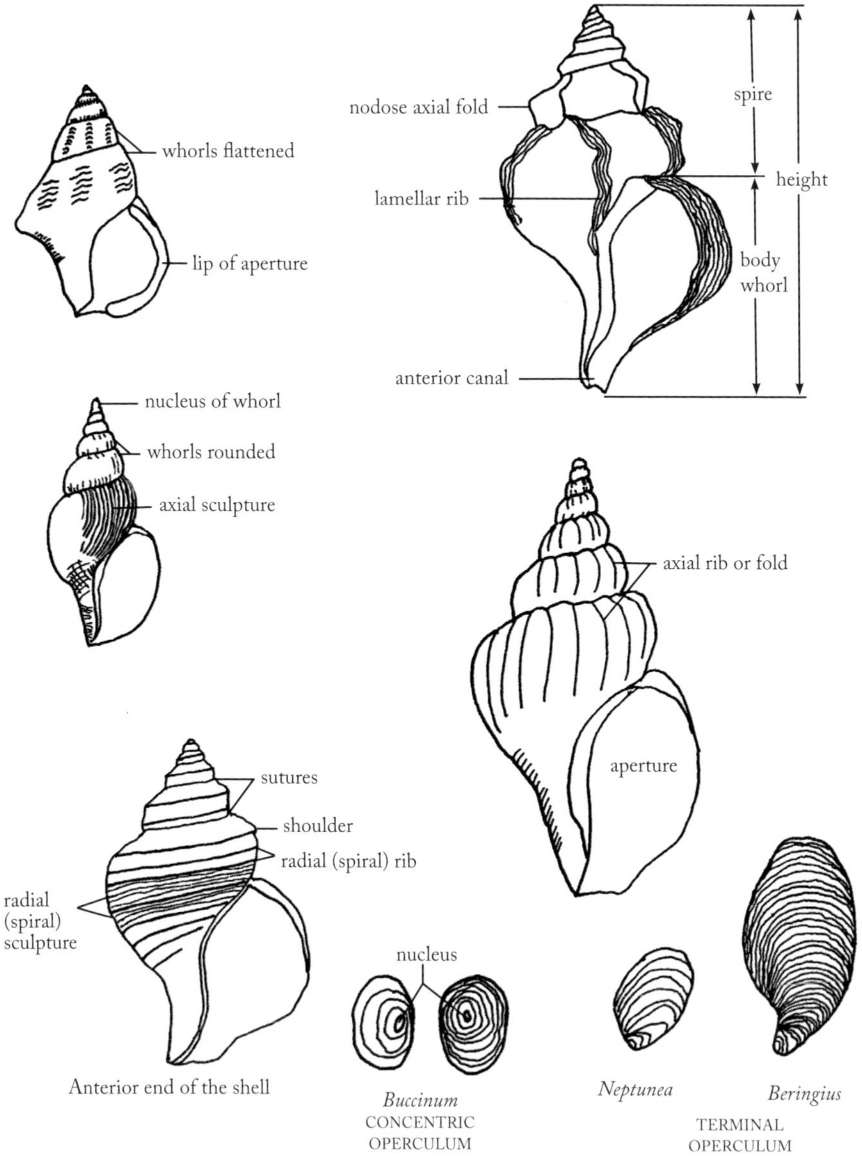

(adapted from MacIntosh 1976)

MOONSNAILS

Pale moonsnail

Euspira pallida

DESCRIPTION: Small spherical shell; umbilicus (pit) very small and slit-like or absent (1). Thin, translucent brown operculum (2). Color white with pale tan periostracum.
SIZE: Height to 5 cm.
DISTRIBUTION: Circumpolar, Arctic Ocean to California. DEPTH: 10-2,000 m.
(Kessler 1985, Clark 2006)

Euspira pallida

Rusty moonsnail

Cryptonatica russa

DESCRIPTION: Small, relatively thick spherical shell. Operculum white, calcareous, and covering aperture; no umbilicus (pit) between aperture and body. Egg case with large capsules arranged in two rows. Uniform rusty brown in color.
SIZE: Height to 4 cm.
DISTRIBUTION: Arctic Ocean through Gulf of Alaska, south to California.
DEPTH: 10-300 m.
(Kessler 1985, Clark 2006)

Cryptonatica russa

Aleutian moonsnail

Cryptonatica aleutica

DESCRIPTION: Small, relatively thick spherical shell. Umbilicus (pit) absent. Operculum white, calcareous, and filling aperture (1). Color reddish-brown on upper ⅔ of shell, white or tan on lower third (2), shell covered with calluses. Juveniles often orange-tan with reddish-brown vertical stripes.
SIZE: Height to 6 cm.
DISTRIBUTION: Bering Sea and Aleutian Islands to California. DEPTH: 1-100 m.
(Clark 2006)

Arctic moonsnail

Cryptonatica affinis

DESCRIPTION: Small, relatively thick spherical shell. Operculum white, calcareous, and covering aperture (1); no umbilicus (pit) between aperture and body (2). Egg case usually less than a complete circle, with raised capsules in a band of 3 to 5 capsules deep. Uniformly gray-white in color.
SIZE: Height to 2.5 cm.
DISTRIBUTION: Circumpolar, Arctic Ocean to California. DEPTH: 1-150 m.
(Kessler 1985, Clark 2006)

SNAILS

Two-keel hairysnail

Trichotropis bicarinata

DESCRIPTION: Medium-sized turbinate shell; aperture about half of shell height. Shell squat, with two prominent spiral keels. Periostracum brownish, with long hairlike tufts on keels (1).
SIZE: Height to 5 cm.
DISTRIBUTION: Arctic Ocean to Bering Sea. DEPTH: 10-475 m.
(Clark 2006)

Cryptonatica aleutica

Cryptonatica affinis

Trichotropis bicarinata

Oregon triton

Fusitriton oregonensis

DESCRIPTION: Large shell covered with bristle-like hairs (1). Radial and axial sculpturing form a pattern of small rectangles. The bristle-like hairs on this shell set it apart from all other shells encountered in its range. Shell white or pink, periostracum green to brown.
SIZE: Height to 15 cm.
DISTRIBUTION: Pribilof Islands, Aleutian Islands, and Bering Sea to California.
DEPTH: 1-200+ m.
(Macintosh 1976, Kessler 1985, Clark 2006)

Frilled dogwinkle

Nucella lamellosa

DESCRIPTION: Small smooth or frilly shell; aperture half or less of shell height. Outer lip broadly flared (1) with 3 rounded teeth. Those found in exposed areas may be smooth; those in sheltered areas may be ornamented with up to a dozen lamellar ribs. Shells vary in color, banding, sculpture, and thickness.
SIZE: Height to 6 cm.
DISTRIBUTION: Bering Sea and Aleutian Islands to California. DEPTH: 0-50 m.
(Harbo 1997, Clark 2006)

Crown trophon

Boreotrophon coronatus

DESCRIPTION: Small spindle-shaped shell; spire less than ⅓ of shell height, with prominent shoulders crowned with strong spines (1). Anterior canal long (2). Color white to pale tan and often stained orange.
SIZE: Height to 6 cm.
DISTRIBUTION: Arctic Ocean and Bering Sea to western Gulf of Alaska.
DEPTH: 40-200 m.
(Kessler 1985, Clark 2006)

Fusitriton oregonensis

Nucella lamellosa

Boreotrophon coronatus

Sinuous whelk

Buccinum plectrum

DESCRIPTION: Shell medium-sized; aperture about ⅓ of shell height. Radial sculpture of fine, well-defined flat-topped ribs (1 mm) set at regular intervals on all whorls. Axial sculpture consists of evenly spaced ribs, about 20 to a whorl. Concentric opercular rings (1). Color pale purple, tan, or gray.
SIZE: Height to 8 cm.
DISTRIBUTION: Chukchi Sea and Bering Sea to Gulf of Alaska. DEPTH: 1-400 m.
(MacIntosh 1976, Clark 2006)

Ladder (silky) whelk

Buccinum scalariforme

DESCRIPTION: Shell medium-sized; aperture about half of shell height. Shell thin, easily broken. Radial sculpture covers entire shell and consists of fine, close striations obvious after close scrutiny. Ribs evenly spaced, rounded, and variable. Concentric opercular growth rings. Color pale gray to tan.
SIMILAR SPECIES: The fine radial striations on *B. scalariforme* vary somewhat in coarseness from shell to shell but should never be confused with the broader flat-topped bands of radial sculpture on *B. plectrum* (above).
SIZE: Height to 8 cm.
DISTRIBUTION: Arctic Ocean, Chukchi Sea, and Bering Sea to Washington. DEPTH: 30-500 m.
(MacIntosh 1976, Clark 2006)

Polar whelk

Buccinum polaris

DESCRIPTION: Shell medium-sized, thin, and fragile. Radial ribs unevenly spaced and of irregular strength. A stronger rib forms the distinct shoulder (1); axial ribs rounded, more prominent on spire than body. Shell covering thin, yellow-brown, and usually peeling off, exposing white shell beneath (2). Concentric opercular rings.
SIMILAR SPECIES: *B. polaris* can be confused with *B. scalariforme* but *B. polaris* has a lighter, thinner shell and the axial sculpture is restricted to the upper whorls and is not at all pronounced on the body whorl. The shell is almost always covered with a thin, dark, bumpy periostracum. Several radial ribs on the body whorl are more pronounced than the others.
SIZE: Height to 7 cm.
DISTRIBUTION: Circumboreal, south to Bering Sea. DEPTH: 30-100 m.
(MacIntosh 1976, Kessler 1985, Clark 2006)

Buccinum plectrum

Buccinum scalariforme

Buccinum polaris

Angular whelk

Buccinum angulosum

DESCRIPTION: Shell medium-sized, aperture about half of shell height. Body whorl distinctive; strongly angled about halfway down the body whorl. Radial sculpture consists of fine, close striations that can be interspersed at regular or irregular intervals with stronger bands. Axial ribs present and regularly spaced; rounded folds restricted to the spire and upper portion of body whorl. Spire whorls tend to be more flattened than in other buccinums. Opercular rings concentric. Color white.

SIMILAR SPECIES: *B. angulosum* may vary in shell shape, but is usually easily separable from other buccinums by the conspicuous angle in the body whorl and the flattened spire whorls.

SIZE: Height to 7 cm.

DISTRIBUTION: Circumboreal, south to Bering Sea. DEPTH: 30-100 m.

(MacIntosh 1976, Clark 2006)

Helmet whelk

Clinopegma magnum

DESCRIPTION: Medium-sized, bulbous shell with wide aperture and short, wide anterior canal. The whorls are strongly shouldered with a single flat rib (1), giving them a flat-sided appearance, though a variation of *C. magnum* without the shoulders has been observed (2). Operculum is dark brown, distinctly wedge-shaped, with a terminal nucleus (3). Color white with hard brown or pale green periostracum.

SIMILAR SPECIES: There are no other common eastern Bering Sea snails with which *C. magnum* can be confused.

SIZE: Average length to 8 cm.

DISTRIBUTION: Bering Sea. DEPTH: 30-100 m.

(MacIntosh 1976, Clark 2006)

Thick-cord whelk

Beringius crebricostatus

DESCRIPTION: Large, spindle-shaped shell; aperture less than half of shell height. Axial sculpture lacking, radial sculpture with few heavy, flat-topped ribs, 3 to 4 above suture, separated by furrows of about the same width. Operculum terminal. Color purple-brown, mauve, or white; aperture white or brown; periostracum thin and yellowish-brown.

SIZE: Height to 13 cm.

DISTRIBUTION: Eastern Aleutian Islands. DEPTH: 3-120 m.

(Clark 2006)

Buccinum angulosum

Clinopegma magnum

Beringius crebricostatus

Kennicott's beringius

Beringius kennicottii

DESCRIPTION: Large inflated, spindle-shaped shell; spire high, without shoulders; anterior canal short. Axial ribs regular, deep, rounded (1). Terminal operculum elongate at nuclear end, large in relation to aperture (2). Color white to tan with thin yellow-brown periostracum. Morphologically variable.
SIZE: Height to 15 cm.
DISTRIBUTION: Central Aleutian Islands to Gulf of Alaska. DEPTH: 1-300 m.
(Kessler 1985, Clark 2006)

"Golden" beringius

Beringius sp.

DESCRIPTION: Medium-sized, spindle-shaped shell; aperture less than half of shell height. Whorls rounded at shoulder; axial sculpture weak to obsolete; radial sculpture faint, if present. Operculum terminal. Color white, with hard lamellar golden periostracum.
SIZE: Height to 13 cm.
DISTRIBUTION: Central Aleutian Islands. DEPTH: 100-200 m.
(Clark 2006)

Bering's beringius

Beringius behringi

DESCRIPTION: Large, inflated spindle-shaped shell; aperture half or more of shell height. Spire whorls are rounded and the spire is quite high. Axial ribs not deep, farther apart on body whorl than spire; radial sculpturing of fine threads on body whorl only. Thin, yellow-brown periostracum flakes off easily but rarely entirely absent. Terminal operculum large, dark brown, elongate, and covered with close-set growth lines throughout its length. Color gray to pale purple, aperture purple.
SIZE: Height to 15 cm.
DISTRIBUTION: Bering Sea, south to eastern Aleutian Islands. DEPTH: 30-150 m.
(MacIntosh 1976, Clark 2006)

Beringius kennicottii

Beringius sp.

Beringius behringi

Northern neptune

Neptunea heros

DESCRIPTION: Large, heavy inflated shell; aperture often nearly ⅔ of shell height. If present, axial ribs may be simple folds or ridges or well developed lamellar ribs. Shoulder nodes usually prominent (1); anterior canal straight. Operculum flame-shaped, with terminal nucleus. Color variable, white, brown, or purple; periostracum very thin, yellowish.
SIZE: Height to 17 cm.
DISTRIBUTION: Arctic Ocean to Bering Sea. DEPTH: 20-70 m.
(Kessler 1985, Clark 2006)

Lyre neptune

Neptunea lyrata

DESCRIPTION: Medium-sized inflated shell; aperture about half or less of shell height. Axial sculpture usually absent; radial ribs prominent, two above the suture, often smaller ribs with fine lirae between. Terminal operculum. Nearly uniform brown tone in color.
SIMILAR SPECIES: *N. pribiloffensis*, *N. heros*, and *N. ventricosa* are all large, abundant, and confusable to one degree or another with *N. lyrata*.
SIZE: Height to 12 cm.
DISTRIBUTION: Eastern Bering Sea, Pribilof Islands, Aleutian Islands, and Gulf of Alaska. DEPTH: 1-100 m.
(Kessler 1985, Clark 2006)

Fat whelk

Neptunea ventricosa

DESCRIPTION: Medium-sized, heavy, squat, and very inflated shell. Radial ribs may or may not be present, usually not prominent on body whorl; axial folds (1) and shoulder nodes usually prominent, but often absent. Spire often with 2 radial ribs of about equal strength; fine radial sculpturing usually absent; lamellar ribs spaced close together near aperture. Aperture wide; anterior canal twisted. Aperture interior occasionally purple; most variable of the neptunes. Operculum flame-shaped, with terminal nucleus. Color brown or purplish tones, rarely white.
SIMILAR SPECIES: *N. ventricosa* is sometimes difficult to distinguish from *N. lyrata* in the non-lamellar form that has strong radial ribs.
SIZE: Height to 12 cm.
DISTRIBUTION: Northeastern Chukchi Sea and Bering Sea to Kodiak Island, Gulf of Alaska, and Cook Inlet. DEPTH: Intertidal, 0-8 m.
(MacIntosh 1976, Kessler 1985, Clark 2006)

Neptunea heros

Neptunea lyrata

Neptunea ventricosa

Lamellar form

Non-lamellar form

Pribilof whelk

Neptunea pribiloffensis

DESCRIPTION: Large, light, thin, often inflated shell. Moderately strong radial ribs on body whorl; axial sculpture absent. Spire with 2 or more radial ribs; shoulder rib strongest; fine, regular radial sculpturing usually present on body whorl and spire. Aperture wide; anterior canal not long; terminal operculum. Color pinkish, white, or tan with dark colored ribs.
SIZE: Height to 15 cm.
DISTRIBUTION: Bering Sea, eastern Aleutian Islands, and Gulf of Alaska to northern California. DEPTH: 30-400 m.
(Kessler 1985, Clark 2006)

Little neptune

Neptunea communis borealis

DESCRIPTION: Small, chunky, spindle-shaped shell. Strong radial ribs; fine, regular radial sculpturing; strong shoulder nodes (1). Aperture elongate, half or more of shell height; anterior canal long; terminal operculum. Color white, pink, brown, or purple.
SIZE: Height to 7 cm.
DISTRIBUTION: Chukchi Sea and Bering Sea. DEPTH: 10-80 m.
(Kessler 1985, Clark 2006)

"Vesteraalen" neptune

Neptunea sp.

DESCRIPTION: Large inflated shell; aperture half to ⅔ of shell height. Axial sculpture lacking; radial sculpture of fine, raised ribs with finer riblets between. Terminal operculum. Color purplish or brown.
SIZE: Height to 15 cm.
DISTRIBUTION: Eastern Aleutian Islands (Islands of Four Mountains). DEPTH: 80-160 m.
(Clark 2006)

Neptunea pribiloffensis

Neptunea communis borealis

Neptunea sp.

Large melon whelk
Pyrulofusus melonis

DESCRIPTION: Large inflated shell, dextral (right-handed) coil; aperture large, elongate. Prominent axial ribs (1); conspicuous radial ribs about 2-mm wide (2). Terminal operculum not elongate at nuclear end, small in relation to aperture (3). Color purple-gray to yellow-brown; aperture purple to orange.
SIZE: Height to 15 cm.
DISTRIBUTION: Bering Sea and eastern Aleutian Islands. DEPTH: 150-400 m.
(Kessler 1985, Clark 2006)

Left-hand melon whelk
Pyrulofusus harpa

DESCRIPTION: Medium-sized inflated shell, sinistrally (left-handed) coiled; aperture large, about ¾ of shell height. Few axial ribs crossed by numerous coarse radial ribs (1). Terminal operculum. Color gray, aperture white, purple, or orange.
SIZE: Height to 13 cm.
DISTRIBUTION: Bering Sea, eastern Aleutian Islands, and Gulf of Alaska.
DEPTH: 1-300 m.
(Kessler 1985, Clark 2006)

Warped whelk
Pyrulofusus deformis

DESCRIPTION: Very large inflated shell, sinistrally coiled (left-handed). Radial sculpturing very fine, threadlike (1); axial sculpturing variable, most often with smoothly rounded ribs more prominent at middle of each whorl. Terminal operculum not elongate at nuclear end, small in relation to aperture. Color gray or yellow-brown, aperture purple to orange.
SIMILAR SPECIES: The only other large, sinistrally coiled snail in the eastern Bering Sea is the closely related *P. harpa*, which is distinguished from *P. deformis* by its coarse and more conspicuous radial ribs.
SIZE: Height to 18 cm.
DISTRIBUTION: Chukchi Sea, eastern Bering Sea, and Gulf of Alaska.
DEPTH: Observed at 176 m, range not known.
(MacIntosh 1976, Kessler 1985, Clark 2006)

Pyrulofusus melonis

Pyrulofusus harpa

Left-handed coil

Pyrulofusus deformis

Left-handed coil

Tulip whelk

Volutopsius middendorffii

DESCRIPTION: Medium-sized inflated shell; aperture about ⅔ of shell height. Axial sculpture lacking; radial sculpture of very fine lirae. The anterior canal is short and quite broad. Characteristic yellowish operculum with terminal nucleus; the inflated nuclear whorls are conspicuously pink. Shell color pink or white, periostracum thin, light brown, when present.
SIZE: Height to 12 cm.
DISTRIBUTION: Bering Sea to Gulf of Alaska. DEPTH: 100-400 m.
(MacIntosh 1976, Clark 2006)

Fragile whelk

Volutopsius fragilis

DESCRIPTION: Medium-sized inflated shell; thin and easily broken. Body whorl long in proportion to total length. Anterior canal short and broad. Radial sculpture consists of fine threads that are often obliterated by wear or encrusting organisms. Axial sculpture consists of shoulder nodes that often continue anteriorly as ribs on body whorl. Terminal operculum; nuclear whorls bulbous and usually conspicuously pink. Color white, often with pink aperture, periostracum thin, brown.
SIZE: Height to 13 cm.
DISTRIBUTION: Northeastern Chukchi Sea and Bering Sea. DEPTH: 30-100 m.
(MacIntosh 1976, Clark 2006)

Frilled whelk

Volutopsius trophonius

DESCRIPTION: Medium-sized inflated shell bearing numerous thin lamellar ribs (1). Aperture large, nearly ⅔ of shell height; terminal operculum. Color pale pinkish, white or light tan.
SIZE: Height to 10 cm.
DISTRIBUTION: Bering Sea, near Pribilof Islands. DEPTH: 150-300 m.
(Clark 2006)

Volutopsius middendorffii

Volutopsius fragilis

Volutopsius trophonius

Volute whelk

Volutopsius castanea

DESCRIPTION: Medium-sized inflated shell. Aperture large, about ⅔ of shell height; terminal operculum. Shell with smoothly angular shoulders that may bear axial wrinkles. Color brown to tan; top of spire with white or purple-gray nuclear whorls.
SIZE: Height to 8 cm.
DISTRIBUTION: Bering Sea, Pribilof Islands, Aleutian Islands, and Kodiak Island to Gulf of Alaska. DEPTH: 1-60 m.
(Kessler 1985, Clark 2006)

Threaded whelk

Lussivolutopsius filosus

DESCRIPTION: Small-sized shell. Aperture more than half of shell height, whorls rounded. Spiral sculpturing prominent, no axial sculpturing. Operculum small and terminal. Color pinkish, mauve, or white, aperture purple-brown.
SIZE: Height to 7 cm.
DISTRIBUTION: Bering Sea to western Gulf of Alaska. DEPTH: 30-100 m.
(Clark 2006)

Arctic whelk

Colus kroyeri

DESCRIPTION: Medium-sized, spindle-shaped shell. Shell thin and very elongate, aperture less than ⅓ of shell height. Canal long, axial sculpture of numerous rounded ribs; radial sculpture of fine lirae. Ovate operculum with a terminal nucleus. Color purple-gray to white or brownish.
SIMILAR SPECIES: The terminal nucleus, high spire, long straight anterior canal, thin shell, and small size distinguishes it from all common eastern Bering Sea snails; when the anterior canal has broken off, it can resemble *Buccinum scalariforme.*
SIZE: Height to 13 cm.
DISTRIBUTION: Circumboreal, Beaufort Sea, Chukchi Sea, south to Bering Sea. DEPTH: 30-80 m.
(MacIntosh 1976, Clark 2006)

Volutopsius castanea

Lussivolutopsius filosus

Colus kroyeri

Oblique whelk

Colus hypolispus

DESCRIPTION: Small, stout, moderately inflated, fusiform shell. Aperture less than half of shell height. Axial sculpture absent; radial sculpture present as a few weak, spaced cords, including some oblique cords near the suture. Terminal operculum. Color pinkish-tan to white with brown periostracum.
SIZE: Height to 6 cm.
DISTRIBUTION: Bering Sea and Prince William Sound, Alaska. DEPTH: 30-150 m.
(Clark 2006)

Shrew whelk

Colus hallii

DESCRIPTION: Small inflated shell; aperture much less than half of shell height. Axial sculpture absent; radial sculpture of fine striae. Terminal operculum. Color white, periostracum brown.
SIZE: Height to 5 cm.
DISTRIBUTION: Bering Sea to Washington. DEPTH: 1-400 m.
(Kessler 1985, Clark 2006)

Ribbed volute

Arctomelon tamikoae

DESCRIPTION: Shell very large, relatively thin and strong; no operculum. Sculpturing consists of numerous low axial ribs (1); sutures T-shaped (2); columella with 2 axial folds. Color chalky blue-gray exteriorly, glossy purple-brown or gray interiorly.
SIZE: Height to 23 cm.
DISTRIBUTION: Bering Sea and Aleutian Islands to Gulf of Alaska.
DEPTH: 100-400 m.
(Clark 2006)

Alaska volute

Arctomelon stearnsii

DESCRIPTION: Very large smooth shell, sutures pinched (1), aperture elongate; no operculum. Columella with 2 prominent axial folds (2). Color chalky blue-gray exteriorly, glossy purple-brown interiorly.
SIZE: Height to 17 cm.
DISTRIBUTION: Chukchi Sea, Bering Sea, St. Lawrence Island, Prince William Sound, southeastern Alaska, south to Washington. DEPTH: 80-500+ m.
(Kessler 1985, Clark 2006)

Colus hypolispus

Colus hallii

Arctomelon tamikoae

Arctomelon stearnsii

SNAIL EGGS

Fat neptune eggs
Neptunea ventricosa eggs

DESCRIPTION: Corncob-shaped, large-diameter clusters (about 3 cm); horizontally positioned capsules.
(Kessler 1985)

Lyre neptune eggs *Neptunea lyrata* eggs
Northern neptune eggs *Neptunea heros* eggs

DESCRIPTION: Single capsules individually attached to shells or substrate but not to one another; arranged in a compact, single-layer cluster. Eggs of these 2 species are not easily distinguished from each other.
(Kessler 1985)

Buccinum eggs
Buccinum sp. eggs

DESCRIPTION: Popcorn-shaped egg clusters; moderate to very large in size. Individual capsules laid on top of one another.
(Kessler 1985)

Beringius eggs
Beringius sp. eggs

DESCRIPTION: Yellow, disk-shaped capsules individually attached to the substrate, well spaced and arranged in a linear series.
(Kessler 1985)

Oregon triton eggs
Fusitriton oregonensis eggs

DESCRIPTION: Coiled clusters laid one layer thick on substrate. Individual capsules flat-topped and shaped like elongate kernels of corn.
(Kessler 1985)

Neptunea ventricosa eggs

Neptunea lyrata
or *Neptunea heros* eggs

Buccinum sp. eggs

Beringius sp. eggs

Fusitriton oregonensis eggs

NUDIBRANCHS

Nudibranchs are very diverse in body, form, and coloration. A guide to the unique nomenclature and keys is in Behrens 1991.

Pacific sea-lemon

Anisodoris nobilis

DESCRIPTION: Large nudibranch covered with conspicuous round tubercles; branchial plume white (1). Color yellow to orange; blackish blotches on the dorsum only.
SIZE: Length to 20 cm.
DISTRIBUTION: Gulf of Alaska to Mexico. DEPTH: 0-100 m.
(Behrens 1991, Clark 2006)

"Spotted" diaulula

Diaulula sp.

DESCRIPTION: Small velvety nudibranch. Body covered with minute tubercles. Color cream-white with numerous small brown spots.
SIZE: Length to 7 cm.
DISTRIBUTION: Aleutian Islands to Gulf of Alaska. DEPTH: 50-100 m.
(Behrens 1991, Clark 2006)

Giant orange tochui

Tochuina tetraquetra

DESCRIPTION: Very large tuberculose nudibranch. Branchial plumes form an irregular series of low, white tufts along the body margins (1). Rhinophores (tentacles) retract into long sheaths (2). Six to ten short, vertical plumose processes encircle shaft below tip. Deep orange-yellow body covered with tubercles tipped with white; foot is salmon pink to yellow. A white band marks the margin of the dorsum.
SIZE: Length to 30 cm.
DISTRIBUTION: Kuril and Aleutian Islands to California. DEPTH: 30-80 m.
(Behrens 1991, Clark 2006)

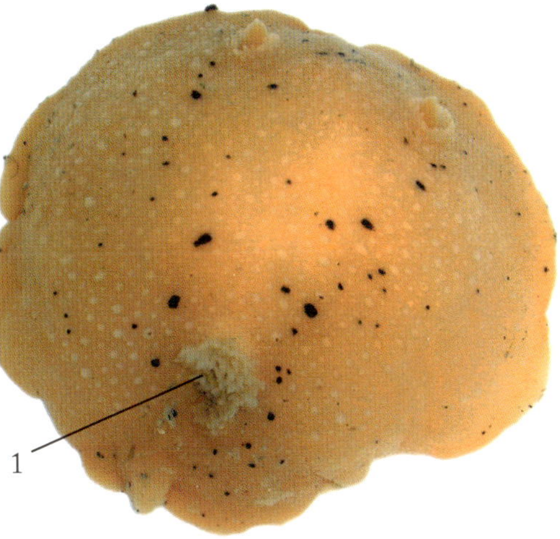

Anisodoris nobilis

Diaulula sp.

Tochuina tetraquetra

Tritonid nudibranch
Tritonia sp.

DESCRIPTION: Large slender nudibranch; body of pictured *Tritonia* is longer and more slender than many others in the genus. Branchial plumes very small, along margin. Color pale orange, branchial plumes white.
SIZE: Length to 10 cm.
DISTRIBUTION: Aleutian Islands. DEPTH: 50-200 m.
(Clark 2006)

Sea-clown triopha
Triopha catalinae

DESCRIPTION: Small, brightly colored nudibranch; body elongate with sparse tubercles. Color white, with orange tubercles, rhinophores, and tips of branchial plumes.
SIZE: Length to 7 cm.
DISTRIBUTION: Aleutian Islands to Baja California. DEPTH: 10-200 m.
(Behrens 1991, Clark 2006)

Dall's dendronotid
Dendronotus dalli

DESCRIPTION: Small nudibranch; body plump. Foot width ⅛ of length. Six or 7 short branched gill tufts; frontal veil bears 4 or 5 extensively branched processes. Rhinophores short (1); crown of rhinophore encircled by 5 processes (2). Background color white to pink; cerata and processes are tipped in white (3).
SIZE: Length to 6 cm.
DISTRIBUTION: Bering Sea, Aleutian Islands to Washington. DEPTH: 50-150 m.
(Behrens 1991, Clark 2006)

Tritonia sp.

Triopha catalinae

Dendronotus dalli

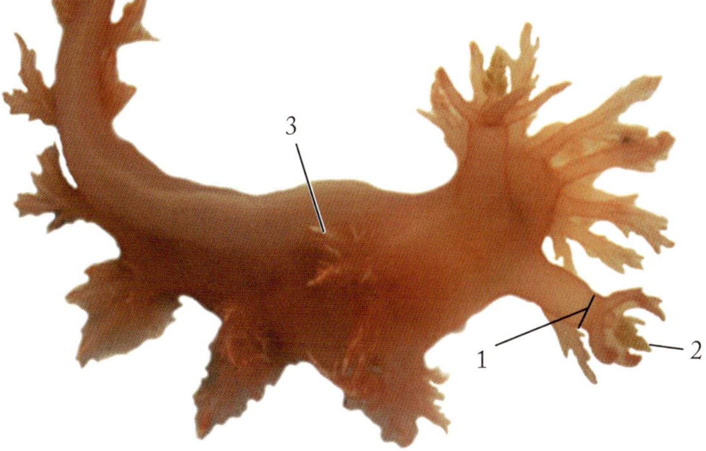

Red frond-aeolis

Dendronotus rufus

DESCRIPTION: Medium-sized nudibranch; foot width 1/6 of length. Lateral rhinophores and frontal veil long and extensively branched; 7-8 ceratal pairs. Color white with red to magenta dots; appendages are tipped with red or magenta; foot edged in red or magenta.

SIZE: Average length 15 cm.

DISTRIBUTION: Kodiak, Alaska, to Washington. DEPTH: No depth information available.

(Behrens 1991)

Lion nudibranch

Melibe leonina

DESCRIPTION: Small nudibranch, large oral hood with 2 rows of cirri along its margin; sail-like rhinophores on hood. Cerata large, flat, and disc-shaped; interior vascular network visible (1). Color yellow-brown to greenish-brown.

SIZE: Length to 9 cm.

DISTRIBUTION: Kodiak Island, Alaska, to Baja California. DEPTH: 0-280 m. Found primarily in eelgrass and kelp beds.

(Behrens 1991)

Golden dirona

Dirona aurantia

DESCRIPTION: Medium-sized ornate nudibranch; body covered with large pointed cerata (1). Color orange to red-orange, tips of cerata white.

SIZE: Length to 12 cm.

DISTRIBUTION: Bering Sea to Washington. DEPTH: 10-100 m.

(Behrens 1991, Clark 2006)

Dendronotus rufus

Melibe leonina

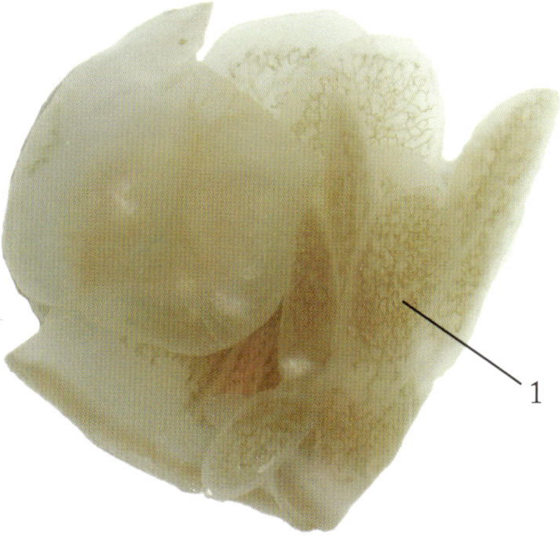

Dirona aurantia

Predaceous aeolis

Flabellina trophina

DESCRIPTION: Body slender, pointed snout, and a long, thick head. Body is translucent white with brownish-red cerata. Head tentacles, rhinophores, and tips of cerata white (1).
SIZE: Length to 5 cm.
DISTRIBUTION: Gulf of Alaska to Oregon. DEPTH: Observed at 13 m, depth range unknown.
(Behrens 1991)

Shag-rug aeolis

Aeolidia papillosa

DESCRIPTION: Medium-sized nudibranch; body tapers from broad head to pointed tail. Large triangular area of white on the head in front of the rhinophores (1). Cerata flattened, lace-shaped, and tightly grouped. Body white to drab gray; rhinophores darker.
SIZE: Length to 10 cm.
DISTRIBUTION: Worldwide in cold and temperate seas. DEPTH: At or below tidal. Found on rocks and in mudflats, also common on pilings and docks.
(Barr and Barr 1983, Behrens 1991)

MUSSELS

Northern horsemussel

Modiolus modiolus

DESCRIPTION: Large rhomboidal mussel; valves relatively thick, inflated, about 3 times as long as wide. Beak of genus *Modiolus* not terminal (1); no radial sculpturing. Wide band of hair on dorsal surface (2). Color pink-lavender with brown periostracum.
SIZE: Length to 18 cm.
DISTRIBUTION: Circumboreal, south to Monterey, California. DEPTH: 0-200 m.
(Kessler 1985, Clark 2006)

Flabellina trophina

Aeolidia papillosa

Modiolus modiolus

Black mussel

Musculus niger

DESCRIPTION: Medium-sized ovate mussel; valves thin, moderately inflated, a little more than twice as long as wide, and bearing fine radial sculpturing. No hair on dorsal surface. Beak of genus *Musculus* not terminal (1). Color dark brown to black.
SIZE: Length to 15 cm.
DISTRIBUTION: Circumboreal, south to Washington. DEPTH: 15-150 m.
(Kessler 1985, Clark 2006)

Discordant mussel

Musculus discors

DISCRIPTION: Small oval mussel; valves thin, inflated, about twice as long as wide. Beak not terminal; anterior sculpturing only; no hair on dorsal surface; often found in a cocoon of soft hairlike fibers (1). Color white, periostracum green to dark brown.
SIZE: Length to 5 cm.
DISTRIBUTION: Circumboreal, south to Washington. DEPTH: 5-150 m.
(Kessler 1985, Clark 2006)

Foolish mussel

Mytilus trossulus

DESCRIPTION: Thin elongate shell, anterior pointed and posterior rounded; smooth exterior with fine concentric lines. Beak of genus *Mytilus* terminal (1). Shiny blue-black or brownish color. *M. edulis* is similar to *M. trossulus*; however, *M. edulis* is almost exclusively found on the Atlantic Coast.
SIZE: Length to 10 cm.
DISTRIBUTION: Arctic to Alaska and south to California. Found intertidally in protected waters on rocks and pilings.
(Kessler 1985, Harbo 1997)

Musculus niger

Musculus niger

Musculus discors

Mytilus trossulus

OYSTER

Pacific oyster

Crassostrea gigas

DESCRIPTION: Shell large and irregularly shaped, sometimes frilly. The lower (left) valve is cupped; the upper (right) valve is flat. Exterior is gray-white, sometimes with purplish markings; interior is polished white or iridescent green. It was introduced to the northwest as seed from Japan in the early 1900s and is raised commercially in Kachemak Bay, Prince William Sound, Annette Island, and other Alaska locations. *C. gigas* does not survive well in northern waters.
SIZE: Length to 60 cm.
DISTRIBUTION: Gulf of Alaska to California.
DEPTH: Intertidal and shallow subtidal zone to 3 m.
(Foster 1991, Harbo 1997)

JINGLE

Alaska (green false) jingle

Pododesmus macroschisma

DESCRIPTION: Large flattened shell of variable shape. Inner surface pearly green or blue-green (1); exterior color white to brown. One or 2 muscle scars. Distinguished by the hole in lower shell (2) where it attaches to substrate.
SIZE: Length to 12 cm.
DISTRIBUTION: Arctic Ocean, Bering Sea, Pribilof Islands, Aleutian Islands, and Gulf of Alaska to Mexico. DEPTH: 0-90 m.
(Kessler 1985, Coan et al. 2000, Clark 2006)

SCALLOPS

Reddish scallop

Chlamys rubida

DESCRIPTION: Large, relatively broad scallop; valves about equal, or longer than wide especially in younger specimens; upper valve with numerous fine, scaly radial riblets. Color variable, pink, purple, or yellow, and uniform on both valves. Often covered with the scallop sponge *Myxilla incrustans*, which lives only on *Chlamys* scallops.
SIZE: Length to 7 cm.
DISTRIBUTION: Albatross Bank, Kodiak Island, east to Kenai Peninsula, Alaska, to California. DEPTH: 1-200 m.
(Harbo 1997, Clark 2006)

Crassostrea gigas

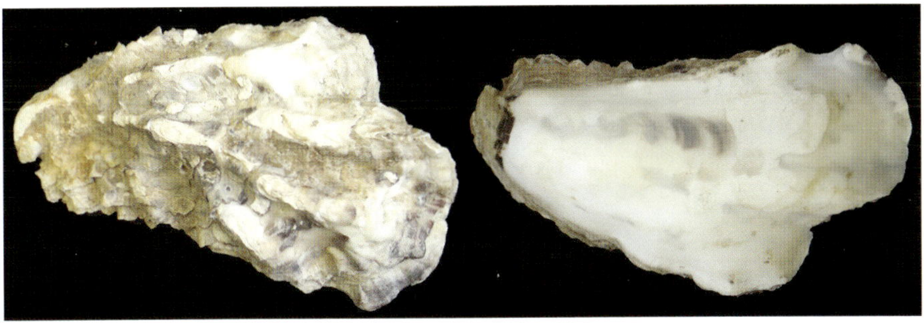

Pododesmus macroschisma

Chlamys rubida

Covered with *Myxilla incrustans*

Weathervane scallop

Patinopectin caurinus

DESCRIPTION: Very large, thin scallop; valves about as long as wide. Ears pronounced and equal (1); strong radial ribs (2). Color of left shell dark brown, convex; right shell white, golden-yellow, or light brown.
SIZE: Length to 25 cm.
DISTRIBUTION: Southeastern Bering Sea, Aleutian Islands and Gulf of Alaska, south to California. DEPTH: 10-200 m.
(Kessler 1985, Coan et al. 2000, Clark 2006)

Patinopectin caurinus

OTHER SCALLOP SPECIES. The giant rock scallop *Crassadoma gigantea* has a heavy, irregularly shaped shell characterized by a purple stain in the interior hinge area. The spiny scallop *Chlamys hastata* has up to 20 large, angular primary ribs topped with prominent spines; the right valve is light pink to white, the left valve is orange red. The island scallop *Chlamys islandica* has a thick, inflated shell, with at least 20 low, flat-topped ribs on both valves; the right valve is white to yellowish, the left valve is dark reddish purple.

EXTERNAL FEATURES OF COCKLES AND CLAMS

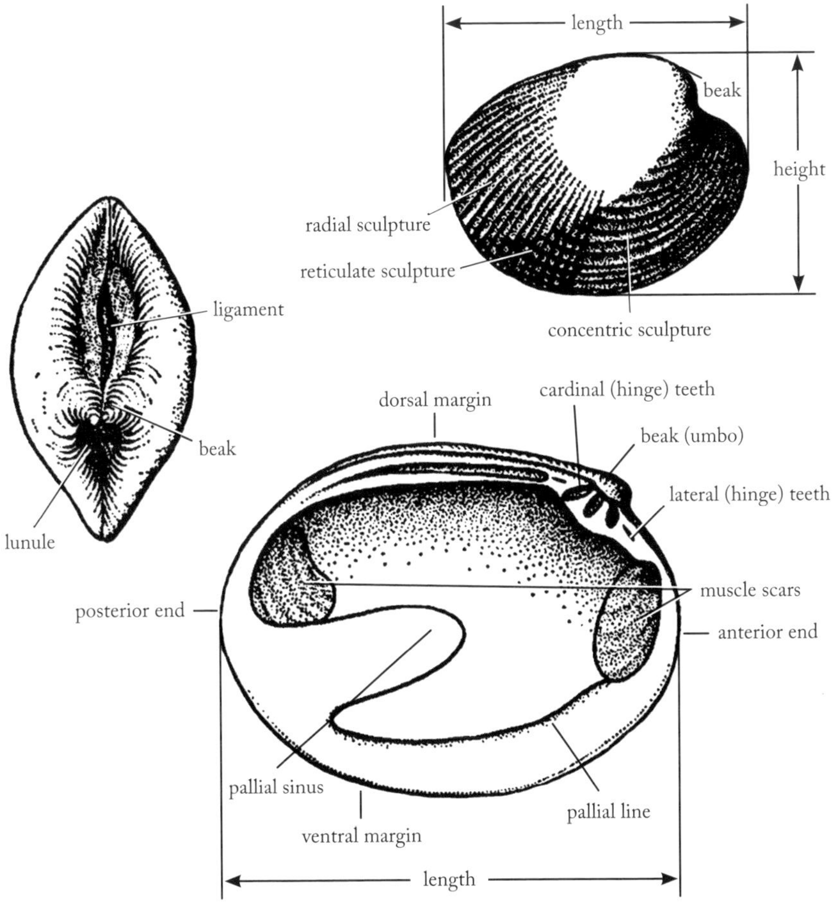

(Adapted from Morris 1966)

COCKLES

Hairy cockle

Clinocardium ciliatum

DESCRIPTION: Medium-sized cockle; valves inflated, slightly oblong in shape, a little longer than high. Tiny hairlike setae in a row along radial rib tops; 35-40 triangular radial ribs. Color white with light brown periostracum.
SIZE: Width to 7 cm.
DISTRIBUTION: Circumboreal, Beaufort Sea, Bering Sea, and Aleutian Islands south to Vancouver Island, British Columbia. DEPTH: 10-150 m.
(Kessler 1985, Coan et al. 2000, Clark 2006)

Nuttall cockle

Clinocardium nuttallii

DESCRIPTION: Large, thick cockle; valves inflated. Adults tending to be higher than long; juveniles are almost round. Strong radial ribs without hairlike setae; commonly with 33-37 coarse radial ribs, each topped with crescent-shaped nodes (1). Color yellow-cream with reddish-brown mottling.
SIZE: Width to 14 cm.
DISTRIBUTION: Punuk Islands, Yukon Delta, west to Unalaska Island, Gulf of Alaska, south to California. DEPTH: Intertidal zone to 30 m.
(Foster 1991, Coan et al. 2000, Clark 2006)

Greenland smoothcockle

Serripes groenlandicus

DESCRIPTION: Large, smooth cockle; valves inflated, about as long as wide. Faint radial sculpture on lateral margins. Beaks straight to slightly off center (1). Exterior of shell is cream with reddish brown flammules as juveniles, becoming dull gray-brown with faint markings as adults.
SIZE: Width to 11 cm.
DISTRIBUTION: Circumboreal, Chukchi Sea, Bering Sea shelf, to Amchitka Island, Aleutian Islands, south to California. DEPTH: 0-80 m.
(Foster 1991, Coan et al. 2000, Clark 2006)

Clinocardium ciliatum

Clinocardium nuttallii

Serripes groenlandicus

Adult

Juvenile

Oblique smoothcockle

Serripes notabilis

DESCRIPTION: Medium-sized smooth cockle; valves inflated, about as long as wide. Faint radial sculpture on lateral margins. Beak directed strongly off center. Color tan with large reddish-brown chevron markings.
SIZE: Width to 9 cm.
DISTRIBUTION: Bering Sea, Pribilof Islands, to northern Gulf of Alaska.
DEPTH: 50-215 m.
(Coan et al. 2000, Clark 2006)

Broad smoothcockle (broad cockle)

Serripes laperousii

DESCRIPTION: Adults medium-sized, thick-shelled, heavy, oblong in shape, two gapes below hinge. Closely resembles *S. groenlandicus*, best distinguishing characteristic is oblong shape of small shells and umbos of larger shells; other characteristics variable. Markings sometimes present on small shells, not zigzagging. Color yellow-tan, with reddish-brown chevron markings.
SIZE: Width to 10 cm.
DISTRIBUTION: Chukchi Sea and Bering Sea, to southeastern Alaska.
DEPTH: 2-80 m.
(Kessler 1985, Coan et al. 2000, Clark 2006)

CLAMS

Washington butterclam

Saxidomus gigantea

DESCRIPTION: Large, thick clam; valves moderately inflated. Concentric ridges (1); no radial ribs. Pallial line unbroken, with a sinus (2); 3 cardinal teeth; lateral teeth present.
SIZE: Width to 13 cm.
DISTRIBUTION: Bering Sea, Seward Peninsula, St. Lawrence Island, to Kodiak Island, Alaska, south to California. DEPTH: 0-40 m.
(Kessler 1985, Coan et al. 2000, Clark 2006)

Serripes notabilis

Serripes laperousii

Saxidomus gigantea

Pacific littleneck

Protothaca staminea

DESCRIPTION: Medium-sized clam, valves moderately inflated. Concentric ridges (1) and radial ribs (2). Pallial line unbroken, with a sinus; 3 cardinal teeth, lateral teeth absent.
SIZE: Length to 8 cm.
DISTRIBUTION: Aleutian Islands, Gulf of Alaska south to Baja California.
DEPTH: 0-10 m.
(Kessler 1985, Coan et al. 2000, Clark 2006)

Yoldia

Yoldia sp.

DESCRIPTION: Comblike hinge teeth. Shell slightly gaping, smooth, glistening, thin, and fragile. Valves ax- or wedge-shaped; color ranges from olive green to chestnut brown.
SIZE: Length to 6.5 cm.
DISTRIBUTION: Circumboreal, south to California. Yoldias are all mud-burrowers. DEPTH: Range unknown.
(Kessler 1985)

Broad yoldia

Megayoldia thraciaeformis

DESCRIPTION: Medium-sized clam. Valves hatchet-shaped. Comblike hinge teeth; spoonlike horizontal hinge projection; shell slightly gaping. Color sometimes brown and dull; outer shell covering flaky when dry; often found in Pacific cod stomachs.
SIZE: Length to 6.5 cm.
DISTRIBUTION: Circumboreal, northern Bering Sea south to California.
DEPTH: 25-760 m.
(Kessler 1985, Coan et al. 2000, Clark 2006)

OTHER SPECIES OF *YOLDIA.* Twelve species of *Yoldia* are found in Alaska waters. Most are long and thin, olive green in color and difficult to tell apart. For more information, see Foster (1991) and Coan et al. (2000).

Protothaca staminea

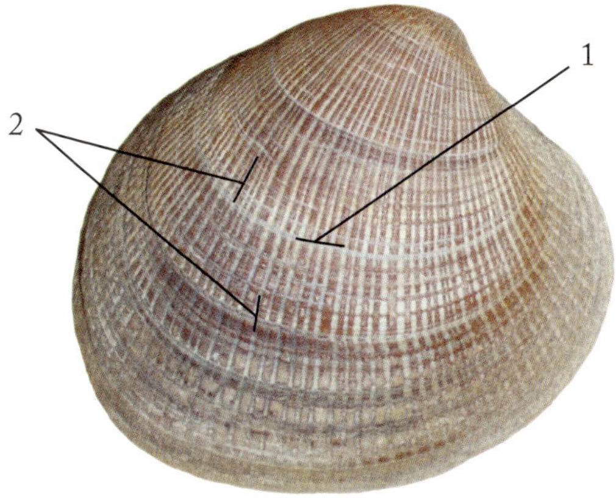

Yoldia sp.

Megayoldia thraciaeformis

Disjunct cleft clam

Conchocele disjuncta

DESCRIPTION: Small subquadrate, chalky clam. No hinged teeth; dorsal groove present (1). Beak anterior (2); anterior edge convex and straight. Color white; outer shell covering yellow-brown.
SIZE: Length to 5 cm.
DISTRIBUTION: Alaska Peninsula to Oregon. Specimen in photo collected from Chiniak Bay, Kodiak Island, Alaska. DEPTH: 100-750 m.
(Bernard 1983, Kessler 1985)

Giant cleft clam

Conchocele bisecta

DESCRIPTION: Large subquadrate, chalky clam; valves thick. No hinged teeth; dorsal groove present (1). Beak anterior (2); anterior edge concave and straight (3). Color whitish. According to Bernard (1972), *C. disjuncta* and *C. bisecta* are mostly distinguished by size; differentiation of the two species has not been confirmed.
SIZE: Length to 11 cm.
DISTRIBUTION: Bering Sea, from Pribilof Islands south to California.
DEPTH: 50-750 m.
(Kessler 1985, Clark 2006)

OTHER CLEFT CLAMS. At least four additional cleft clams are found in Alaska; usually small, they are not commonly found in trawls or pots. For more information see Coan et al. (2000).

Elegant soft shell

Mya elegans

DESCRIPTION: Large quadrate clam. Spoonlike horizontal hinge projection (1) fits into opposite socket. Sinus on left shell extending well beyond umbo (2); posterior of shell narrowly rounded, anterior broadly rounded. Color whitish.
SIZE: Length to 14 cm.
DISTRIBUTION: Pribilof Islands to south of Alaska Peninsula. DEPTH: 0-100 m.
(Kessler 1985)

Conchocele disjuncta

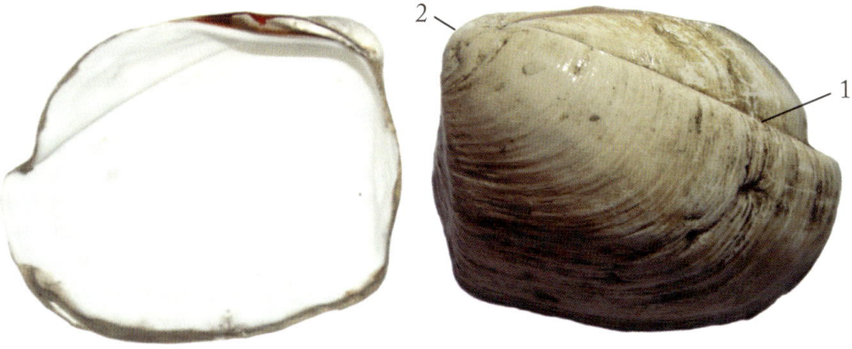

Conchocele bisecta

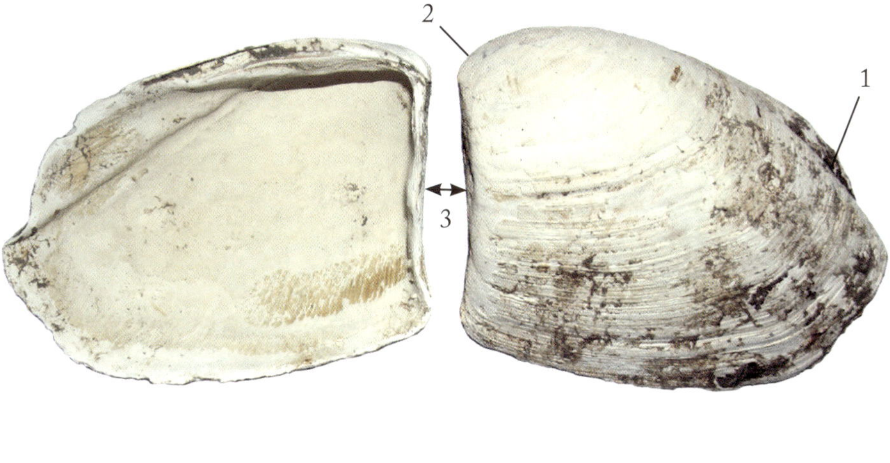

Mya elegans

Alaska great-tellin
Tellina lutea

DESCRIPTION: Large, heavy, subovate clam. Valves moderately inflated; longer posteriorly; posterior end slightly twisted. Pallial line unbroken, with a sinus; 2 cardinal teeth (1). Shell interior pink-colored; outer shell white, chalky; periostracum glossy green in young, brown in adults.
SIZE: Length to 13 cm.
DISTRIBUTION: Arctic Ocean, Chukchi Sea, Bering Sea, Pribilof Islands to Cook Inlet, Alaska. DEPTH: 0-100 m.
(Kessler 1985, Coan et al 2000, Clark 2006)

Arctic roughmya
Panomya norvegica

DESCRIPTION: Large, thick quadrate clam; valves heavy, anterior end rounded, posterior end truncated. Conspicuous pallial scar line (1); single long, pointed hinge tooth on right shell (2); external hinge ligament present. Color white, periostracum thin, brown, and flaky.
SIZE: Length to 11 cm.
DISTRIBUTION: Circumpolar, Bering Sea, south to Oregon. DEPTH: 20-640 m.
(Kessler 1985, Clark 2006)

Arctic hiatella
Hiatella arctica

DESCRIPTION: Small, usually distorted, elongate clam. Inconspicuous pallial scar line. Sculpture of irregular growth lines; external hinge ligament present. Color white with thin, flaky brown periostracum.
SIZE: Length to 5 cm.
DISTRIBUTION: Worldwide, possibly spread by human activity; Arctic Ocean, Bering Sea south to western coast of North America to Chile. DEPTH: 0-1,190 m.
(Kessler 1985, Coan et al. 2000, Clark 2006)

Tellina lutea

Panomya norvegica

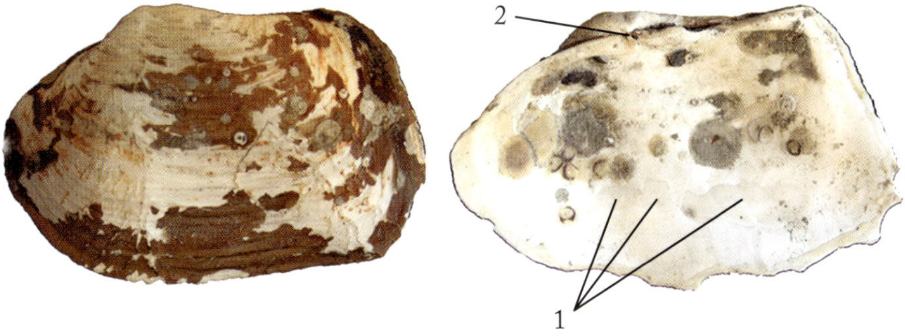

Hiatella arctica

Pilsbry piddock

Zirfaea pilsbryi

DESCRIPTION: Large clam with large body that does not completely fit between protective shells; diagonal groove separates the shell's anterior and posterior portions (1). Rasping anterior end has spines and teeth (2), posterior end has concentric growth lines. Shells gape widely at both ends. Shell white; split siphons rusty-red with white bumps at tip.
SIZE: Length to 15 cm.
DISTRIBUTION: Chukchi Sea to Baja California. DEPTH: Intertidal zone to 120+ m. Burrows into mud, clay, shale, waterlogged wood, and sandstone.
(Foster 1991, Harbo 1997)

OTHER PIDDOCKS. The flat-tip piddock *Penitella penita* and the chimney piddock *P. hopkinsi* are the two other piddocks found in Alaska waters. These clams are highly modified for digging burrows in solid substrates.

Alaska razor clam

Siliqua alta

DESCRIPTION: Large broad clam; valves thin, broad, about ⅓ as wide as long. Gaping at both ends; internal rib perpendicular (1). Shell heavier, ends less rounded and rib stronger and narrower than *S. patula*. Interior chalky, white or tinted purple; neck and mantle skin dark-colored; periostracum relatively thick, glossy, olive to dark brown.
SIZE: Length to 20 cm.
DISTRIBUTION: Chukchi Sea, Bering Sea, Cook Inlet, Alaska, south to California. DEPTH: 0-85 m.
(Kessler 1985, Coan et al. 2000, Clark 2006)

Pacific razor clam

Siliqua patula

DESCRIPTION: Large broad clam; valves thin, about ⅓ as wide as long and gaping at both ends. Internal rib slants anteriorly (1). Periostracum olive brown, varnish-like; interior glossy, white or tinted purple; neck and mantle skin light-colored.
SIZE: Length to 19 cm.
DISTRIBUTION: Gulf of Alaska to California. DEPTH: 0-55 m.
(Kessler 1985, Clark 2006)

Zirfaea pilsbryi

Siliqua alta

Siliqua patula

Arctic surfclam

Mactromeris polynyma

DESCRIPTION: Large trigonal clam; valves relatively thin, moderately inflated, about ⅓ longer than wide, smooth except for growth lines. Color white with tan to dark-brown periostracum.
SIZE: Length to 15 cm.
DISTRIBUTION: Circumboreal, Bering Sea, Pribilof Islands south to Washington.
DEPTH: 0-110 m.
(Coan et al. 2000, Clark 2006)

Fat gaper clam

Tresus capax

DESCRIPTION: Very large rhomboidal clam; valves inflated with pronounced ventral margin, posterior end broadly gaping. Ligament socket triangular (1). Color white with olive periostracum.
SIZE: Length to 28 cm.
DISTRIBUTION: Gulf of Alaska to California. DEPTH: 0-20 m.
(Kessler 1985, Clark 2006)

Arctic astarte

Astarte arctica

DESCRIPTION: Small, smoothly rounded, glossy shell. Shell convex and compact, shell length and height are comparable in size. Color white with olive to dark-brown periostracum.
SIZE: Width to 3.5 cm.
DISTRIBUTION: Bering Sea, Gulf of Alaska south to Washington.
DEPTH: 20-30 m.
(Coan et al. 2000, Clark 2006)

Mactromeris polynyma

Tresus capax

Astarte arctica

Elliptical astarte

Astarte elliptica

DESCRIPTION: Small rounded, triangular, slightly compressed shell. The ribs are heavy, even, and broader than those of other Alaska *Astarte* species. The beaks (umbos) are prominent. Color white with dark periostracum that may peel away.
SIZE: Width to 5 cm.
DISTRIBUTION: Circumboreal, Bering Sea, Gulf of Alaska south to Washington.
DEPTH: 5-130 m.
(Foster 1991, Coan et al. 2000, Clark 2006)

OTHER CLAMS. There are many other soft shell clams, roughmyas, and astartes found in Alaska waters. For information on species not found in this guide, see Foster (1991) and Coan et al. (2000).

BRACHIOPODS

Black lamp shell

Hemithiris psittacea

DESCRIPTION: Valves triangular, with radial lines; anterior edge of valves with strong downward wave. Color dark gray or black. Brachiopods are attached to substrate by a pedicel (stalk).
SIZE: Length to 2 cm.
DISTRIBUTION: Circumboreal, south to Oregon. DEPTH: 15-400 m.
(Clark 2006)

Transverse lamp shell

Terebratalia transversa

DESCRIPTION: Large variable brachiopod; distinguished by its undulating margin (1). Variable in dimension, may be as long as wide or up to twice as wide as long. May be smooth or radially ribbed. Pedicel very short, stout.
SIZE: Length to 5 cm.
DISTRIBUTION: Aleutian Islands to California. DEPTH: 1-100+ m.
(Kessler 1985, Clark 2006)

OTHER LAMP SHELLS. Several other lamp shells have been reported in Alaska waters. Specimens not fitting the above descriptions should be saved for positive identification.

Astarte elliptica

Hemithiris psittacea

Terebratalia transversa

SQUIDS

Magister armhook squid

Berryteuthis magister

DESCRIPTION: Large-sized body pointed at end of mantle (1); hooks on arms. Body cylindrical; fins triangular and at posterior end of mantle (2); 8 arms, and 2 tentacles that are long and retractable. Color red.
SIZE: Mantle length to 34 cm.
DISTRIBUTION: Bering Sea, eastern Aleutian Islands to southeastern Alaska.
DEPTH: Pelagic, roaming as deep as 750 m near the continental shelf.
(Kessler 1985, Jorgensen 2009)

North Pacific bobtail squid

Rossia pacifica

DESCRIPTION: Small-sized body rounded at end of mantle (1); fins semicircular; mantle free all around. Fins forward of posterior end of mantle (2); 8 arms, and 2 tentacles that are long and retractable. Often red-brown, but changeable in color due to presence of chromatophores.
SIZE: Mantle length to 10 cm.
DISTRIBUTION: North Pacific Rim from Japan to southern California.
DEPTH: Low intertidal to 1,000 m.
(Kessler 1985, Jorgensen 2009)

OCTOPUS

Pacific giant octopus

Enteroctopus dofleini

DESCRIPTION: Body round, without fins; 8 arms that are 3 to 5 times body length. Extensive skin folds; large conical warts above eyes. Skin color reddish on the upper side and paler beneath.
SIZE: Total length to 2.5 m.
DISTRIBUTION: Bering Sea, Aleutian Islands to Korea, and the North Pacific to California. DEPTH: Intertidal to 1,000 m.
(Kessler 1985, Jorgensen 2009)

Berryteuthis magister

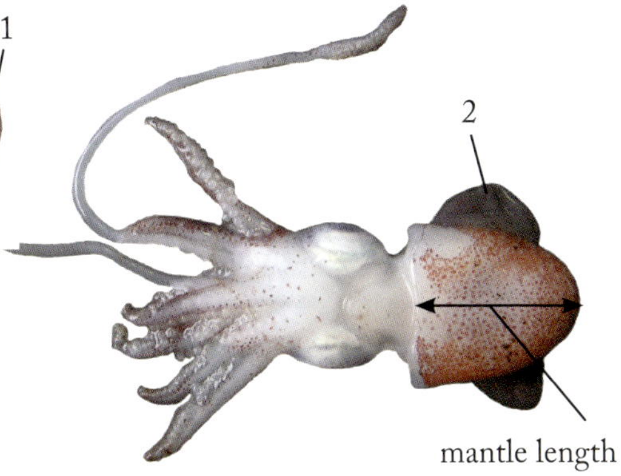

Rossia pacifica

Enteroctopus dofleini

SEA SPIDER

Giant sea spider

Colossendeis sp.

DESCRIPTION: Very large, long-legged pycnogonid. Body and legs long, segments not very distinct, spider-like in appearance. Color red-orange.
SIZE: Diameter to 30 cm.
DISTRIBUTION: Bering Sea to Gulf of Alaska. DEPTH: 200-3,000 m.
(Clark 2006)

BARNACLES

Eared barnacle

Tarasovium cornutum

DESCRIPTION: Small, stalked, laterally compressed barnacle. Calcareous plates small. Keel-like back with 2 ear-shaped projections at stalked end (1). Stalk relatively long and pinkish-tan; plates white.
SIZE: Length to 3 cm.
DISTRIBUTION: Bering Sea to Gulf of Alaska. DEPTH: Below 100 m.
(Kessler 1985, Clark 2006)

Pelagic gooseneck barnacle

Lepas sp.

DESCRIPTION: Small, stalked, laterally compressed barnacle. Calcareous plates large, smooth; peduncle purplish-brown, relatively long; color of plates pale blue to white.
SIZE: Length to 8 cm.
DISTRIBUTION: North Pacific. Pelagic, living in the open sea usually near the surface.
(Clark 2007)

Colossendeis sp.

Tarasovium cornutum

Lepas sp.

Beaked barnacle

Balanus rostratus

DESCRIPTION: Large barnacle. Tips of terga (shell plates) long and pointed, drawn out into a beak-like shape. Six calcareous plates, with rostrum overlapping adjacent plates. Shell wall hollow, exterior plates smooth or with shallow longitudinal striations; overlapping portions of plates usually glossy. Base of barnacle well-calcified.
SIZE: Diameter to 5 cm.
DISTRIBUTION: Pacific Coast, north and south of Alaska Peninsula and off Japan. DEPTH: Mostly subtidal. Sometimes encountered on floating docks and on rocks in tidal streams.
(Kessler 1985, Kozloff 1987)

Giant barnacle

Chirona evermanni

DESCRIPTION: Very large barnacle. Characteristically large, branched cluster; height 1.5 to 2 times width; terga (shell plates) with long, thin spur (1); shell wall solid, terga without tubes.
SIZE: Height to 13 cm.
DISTRIBUTION: Bering Sea to Gulf of Alaska. DEPTH: 100-400 m.
(Kessler 1985, Clark 2006)

Crab barnacle

Hesperibalanus hesperius

DESCRIPTION: Small barnacle. Orifice more or less rectangular, and about ⅓ of shell diameter. Shell wall solid without tubes.
SIZE: Diameter to 1 cm.
DISTRIBUTION: Commonly found on crabs and snails in the Bering Sea, Gulf of Alaska to California.
(Kessler 1985, Arnsberg 2001)

Balanus rostratus

Chirona evermanni

Hesperibalanus hesperius

ISOPODS

Gravel isopod

Synidotea sp.

DESCRIPTION: Small isopod. Pleotelson broad, slightly indented posteriorly (1). Color pale orange-brown.
SIZE: Length to 2.5 cm.
DISTRIBUTION: Aleutian Islands. DEPTH: Below 200 m.
(Clark 2006)

Sea cockroach

Rocinela angustata

DESCRIPTION: Medium-sized isopod. Thoracic segments much larger than abdominal segments. Pleotelson relatively large, tapering, posteriorly rounded (1). Color olive and pink.
SIZE: Length to 4 cm.
DISTRIBUTION: Bering Sea to Gulf of Alaska. DEPTH: 100-300 m.
(Kozloff 1987, Clark 2006)

Spiky arcturid

Arcturus sp.

DESCRIPTION: Medium-sized arcturid isopod. Body rather slender, subcylindrical, with 2 parallel rows of long, stout, spike-like spines on dorsal side of body segments. Antennae very long. Color mottled reddish-brown and cream.
SIZE: Length to 4 cm.
DISTRIBUTION: Bering Sea and Aleutian Islands. DEPTH: 50-350 m.
(Clark 2006)

Synidotea sp.

Rocinela angustata

Arcturus sp.

Spiny arcturid
Arcturus sp.

DESCRIPTION: Small bristly arcturid. Body subcylindrical, bearing numerous bristling spines. Color cream and light brown.
SIZE: Length to 3 cm.
DISTRIBUTION: Aleutian Islands. DEPTH: 200-300 m.
(Clark 2006)

Arcturus sp.

AMPHIPOD

Skeleton shrimp
Caprella sp.

DESCRIPTION: Small, slender, modified amphipod. Similar in appearance to terrestrial "walking stick" insects. Color pale tan, pink, brown, or white.
SIZE: Length to 5 cm.
DISTRIBUTION: Arctic Ocean to Mexico. Several marine species in Alaska waters. DEPTH: 0-1,000+ m.
(Kozloff 1987, Clark 2006)

Caprella sp.

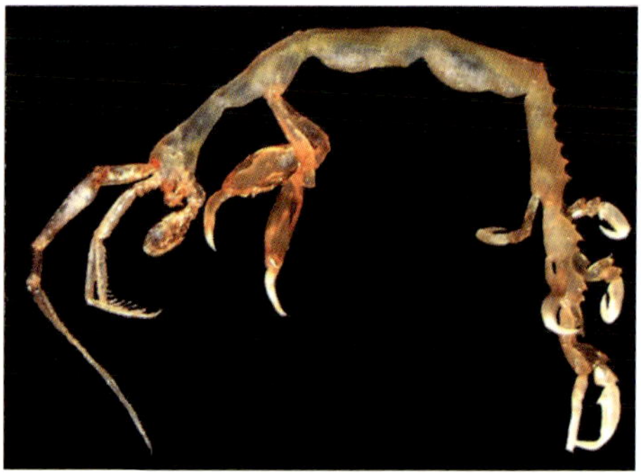

SHRIMPS

SHRIMP ANATOMY

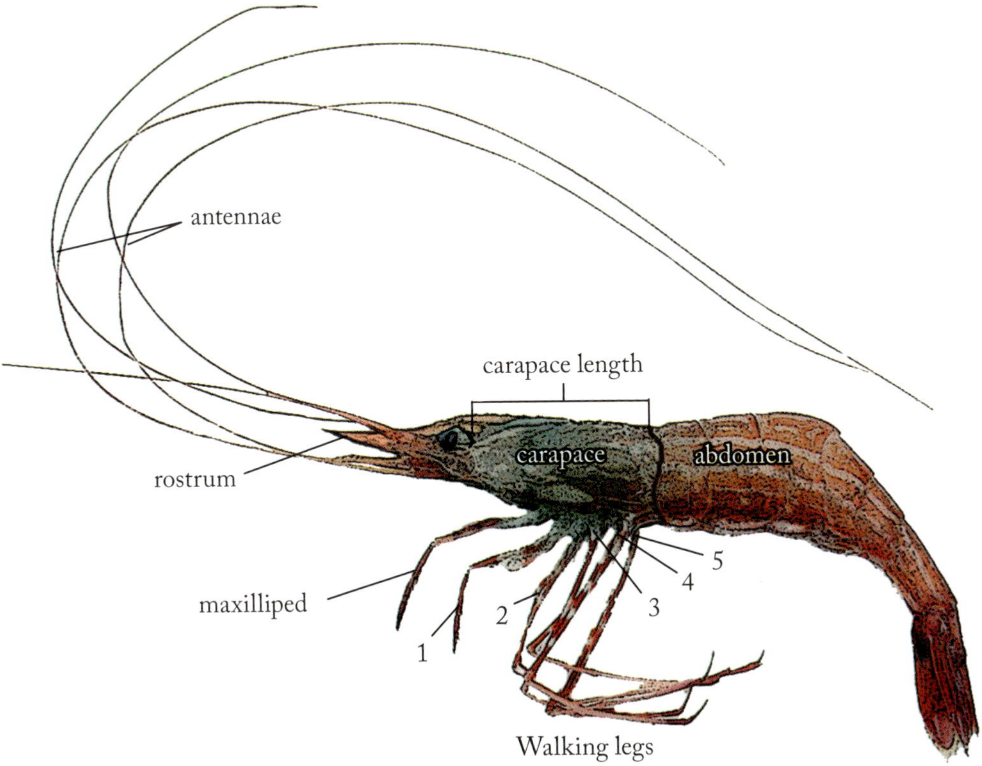

Genus *Pandalopsis*. First walking leg without a claw; maxillipeds not black-tipped; spines jointed to rostrum, separated by fine hairs; 2 pairs of long antennae; base of first walking leg laterally expanded.

Sidestripe shrimp

Pandalopsis dispar

DESCRIPTION: Medium-sized pandalid shrimp. Body slender; base of first walking leg laterally expanded (1). Rostrum twice as long as carapace and bearing spines all along dorsal and ventral sides. Color pale orange with light stripes on each side of abdomen; legs with prominent red and white bands (2).

SIZE: Carapace length to 3 cm.

DISTRIBUTION: Bering Sea, Pribilof Islands to South America. DEPTH: 45-800 m.

(Butler 1980, Kessler 1985, Clark 2006)

Aleutian bigeye

Pandalopsis aleutica

DESCRIPTION: Medium-sized pandalid shrimp. Body slender; base of first walking leg laterally expanded (1). Rostrum about as long as carapace, dorsal side smooth; dorsal spines of carapace relatively large, extending to about midpoint of carapace. Color of carapace pale pink, reddish on dorsal ridge and rostrum; rostrum with distinctive white band near distal end (2); abdomen dark pink with irregular white bands and elongate blotches; legs pink.

SIZE: Carapace length to 3.5 cm.

DISTRIBUTION: Bering Sea and Aleutian Islands. DEPTH: below 400 m.

(Clark 2006)

Genus *Pandalus*. First walking leg without a claw; maxillipeds not black-tipped; spines jointed to rostrum, separated by fine hairs; 1 pair of long antennae; base of first walking leg not laterally expanded.

Spot shrimp (prawn)

Pandalus platyceros

DESCRIPTION: Large pandalid shrimp. Body relatively stout; rostrum longer than carapace, with fine spines. Broad white stripes on carapace (1); banded legs; dorsal spines not extending past middle of carapace (2); white spots on abdomen (3).

SIZE: Carapace length to 6 cm.

DISTRIBUTION: Aleutian Islands to California. DEPTH: 5-500 m.

(Butler 1980, Clark 2006)

Pandalopsis dispar

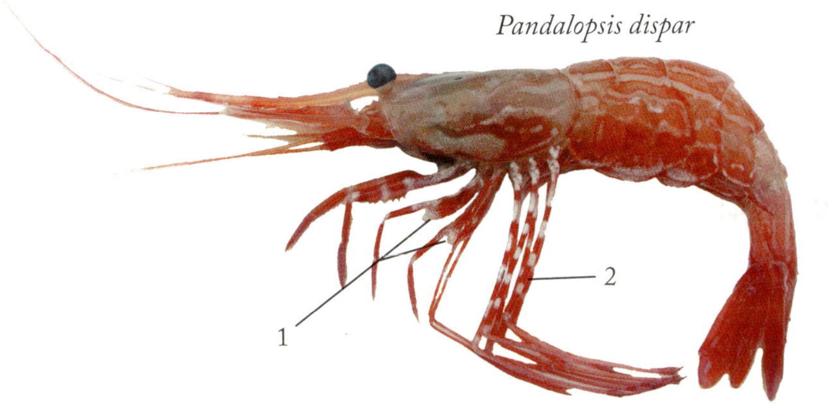

Pandalopsis aleutica

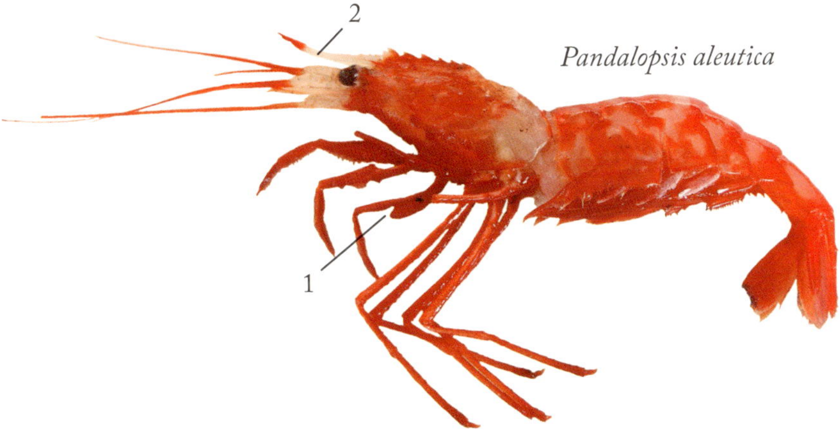

Pandalus platyceros

Coonstripe Shrimp

Pandalus hypsinotus

DESCRIPTION: Large pandalid shrimp. Body stout, rostrum longer than carapace; dorsal spines extending beyond midpoint of carapace. Third abdominal segment lacking keel or spine. Color dark orange-brown tones with white blotches on carapace (1), and maroon or reddish blotches on abdomen (2). Legs orange-brown with maroon and white bands.
SIZE: Carapace length to 4 cm.
DISTRIBUTION: Bering Sea to Washington. DEPTH: 5-460 m.
(Clark 2006)

Dock shrimp

Pandalus danae

DESCRIPTION: Medium-sized pandalid shrimp. Body stout, rostrum longer than carapace; dorsal spines extending past midpoint of carapace. Third abdominal segment lacks keel and spine. Color is subdued brown, rostrum with white band near tip (1), carapace with white and maroon spots (2), abdominal segments with thick, slanted, maroon lines (3); legs white-banded with maroon and yellow.
SIZE: Carapace length to 3 cm.
DISTRIBUTION: Gulf of Alaska to Mexico. DEPTH: 0-185 m.
(Butler 1980, Clark 2006)

Alaskan pink shrimp

Pandalus eous

DESCRIPTION: Medium-sized pandalid shrimp. Body rather slender; rostrum longer than carapace with prominent spines on dorsal and ventral sides. Third abdominal segment with median keel ending in prominent spine forward of rear edge of segment (1); fourth abdominal segment also has a dorsal spine at the posterior margin. Rostrum 1.5-2 times as long as the carapace. Solid translucent pink carapace without distinctive markings, legs not banded. *P. eous* is recognized as a separate species found in the Pacific, whereas *P. borealis* is an Atlantic species.
SIZE: Carapace length to 3 cm.
DISTRIBUTION: Bering Sea to Oregon. DEPTH: 15-1,380 m.
(Butler 1980, Barr and Barr 1983, Squires 1992)

Pandalus hypsinotus

Pandalus danae

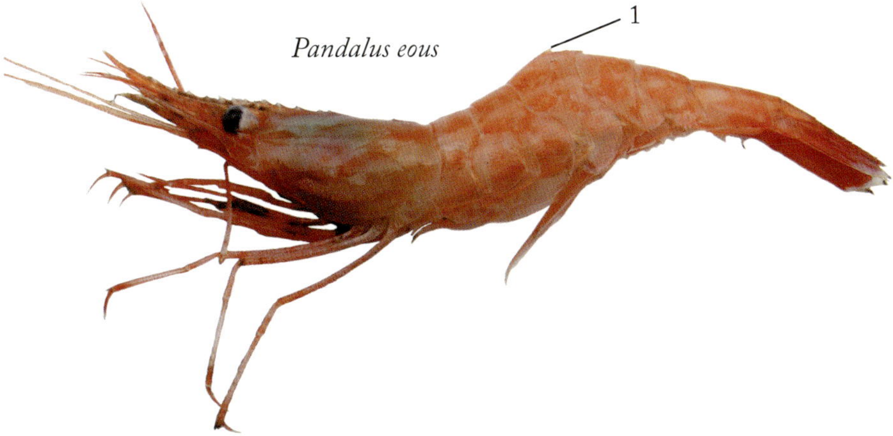

Pandalus eous

Humpy shrimp

Pandalus goniurus

DESCRIPTION: Medium-sized, pandalid shrimp. Body slender; rostrum slender, longer than carapace. Third abdominal segment with median keel ending in a spine (1). Color translucent pale pink to white with slanting red lines on abdomen (2); rostrum with red line crossed by one or two white bands at midpoint.
SIZE: Carapace length to 3 cm.
DISTRIBUTION: Chukchi Sea, Bering Sea to Washington. DEPTH: 3-1,500 m.
(Butler 1980, Clark 2006)

Yellowleg pandalid

Pandalus tridens

DESCRIPTION: Medium-sized pandalid shrimp. Body slender; rostrum longer than carapace. No stripes or spots on carapace or abdomen. Color pale pink with dark pink or reddish mottling; legs with distinctive yellow bands (1).
SIZE: Carapace length to 3 cm.
DISTRIBUTION: Bering Sea to California. DEPTH: 5-2,000 m.
(Butler 1980, Clark 2006)

Spiny lebbeid

Lebbeus groenlandicus

DESCRIPTION: Medium-sized stout-bodied shrimp. Rostrum slender, shorter than carapace. First walking leg with a distinct claw (1); maxillipeds with black tips; rostral spines immovable, one spine above eye. Four large prominent dorsal spines (2); pleura with sharp triangular spine on ventral edge (3). Color orange-brown or cream with reddish-brown bands and mottling.
SIZE: Carapace length to 3 cm.
DISTRIBUTION: Bering Sea to British Columbia. DEPTH: 10-520 m.
(Butler 1980, Kessler 1985, Clark 2006)

OTHER LEBBEIDS. The polar lebbeid *L. polaris* is distinguished from *L. groenlandicus* by its red and orange spotted coloration and three small, low dorsal spines.

Pandalus goniurus

Pandalus tridens

Lebbeus groenlandicus

Color variations

Arctic argid
Argis dentata

DESCRIPTION: Medium-sized stout shrimp. Body slightly flattened; first walking leg with a subchelate claw (1). Eyes small, closely spaced, partially hidden from above, with cone-like tubercles; 2 median dorsal spines on carapace; keels on sixth abdominal segment do not extend past the rear edge of segment and end in a knob or spine (2). Color cream with fine dark speckling and dark chevrons across carapace and abdomen (3).
SIZE: Carapace length to 4 cm.
DISTRIBUTION: Circumboreal, south to Bering Sea. DEPTH: 0-290 m.
(Butler 1980, Kessler 1985, Clark 2006)

OTHER ARGIDS. The northern argid *Argis lar* is distinguished from *A. dentata* by keels on the sixth abdominal segment that extend beyond the posterior margin of the segment. The split-eye argid *A. ovifer* occurs in deep water and has a sharp tubercle on each eyestalk.

Sculptured shrimp
Sclerocrangon boreas

DESCRIPTION: Large, robust, deeply sculptured "tank-like" shrimp. First walking leg with a subchelate claw (1). Eyes not small, not hidden from above; antennal scales wide (2). Wide body; square-shaped carapace about one-fifth longer than wide; 3 median dorsal spines on carapace (3); rostrum tip horizontal. Color mottled brown and white patterns.
SIZE: Carapace length to 4 cm.
DISTRIBUTION: Bering Sea to Washington. DEPTH: 0-370 m.
(Butler 1980, Kessler 1985)

OTHER SHRIMPS. Many more species of shrimps are commonly found in Alaska waters than are described in this guide. Within the large, diverse hippolytid family, there are at least nine *Eualus* and eleven *Heptacarpus* shrimp species commonly found in Alaska waters. In addition, commonly found shrimps in the crangonid family include at least nine *Crangon* species and five argid species. Butler (1980) and Jensen (1995) have keys and characteristics of these and other shrimps found in Alaska waters.

Argis dentata

Sclerocrangon boreas

CRABS

EXTERNAL FEATURES OF CRABS

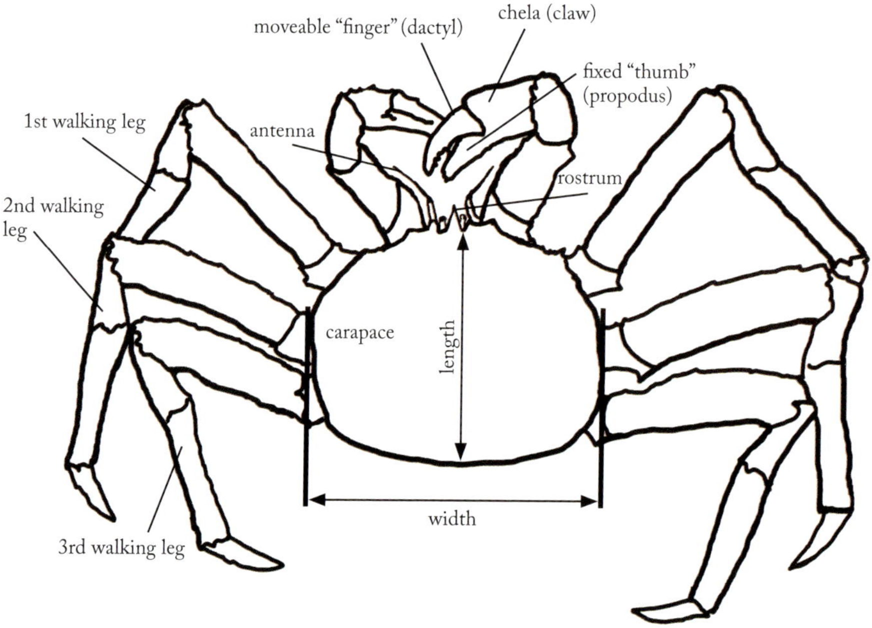

Puget Sound king crab

Lopholithodes mandtii

DESCRIPTION: Large rock-like crab. Legs short, stubby, and fitting tightly against body; no hole formed by the margin of first and second legs. Carapace heavy with 4 large, cone-shaped humps (1-4). Coloration of adults is orange-brown with purple markings.
SIZE: Carapace length to 23 cm.
DISTRIBUTION: Gulf of Alaska to California. DEPTH: Shallow subtidal to 100 m.
(Donaldson and Byersdorfer 2005)

Brown box crab

Lopholithodes foraminatus

DESCRIPTION: Large rock-like crab. Legs short, stubby, and fitting tightly against body; hole is formed on each side of the crab by the joining indentations (1). Predominantly drab reddish-brown or tan color.
SIZE: Carapace length to 18.5 cm.
DISTRIBUTION: Gulf of Alaska to California. DEPTH: Low intertidal to 547 m.
(Donaldson and Byersdorfer 2005)

Lopholithodes mandtii

(Preserved)

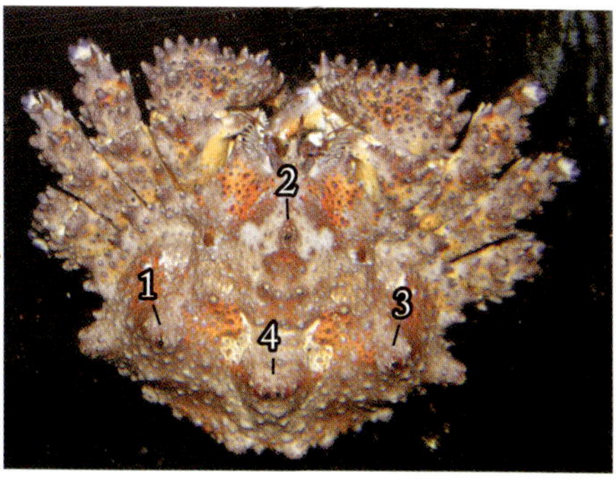

Lopholithodes foraminatus

Rhinoceros crab

Rhinolithodes wosnessenskii

DESCRIPTION: Small triangular-shaped crab. Carapace with deep semicircular depression around a smooth knob (1); remainder of carapace covered with tubercles. Legs short and stout, with pointed spines with coarse tipped hairs. Carapace light brown, base of rostrum orange, depressed area orange and white.
SIZE: Carapace length to 6 cm.
DISTRIBUTION: Aleutian Islands to California, on rock, gravel, or shell bottoms.
DEPTH: 6-102 m.
(Donaldson and Byersdorfer 2005)

Heart crab

Phyllolithodes papillosus

DESCRIPTION: Medium-sized triangular-shaped crab. Carapace triangular in outline with raised heart-shaped sculpturing. Claws, legs, and edge of carapace armed with long flattened projections (1). Color grayish or brown, juveniles often white with purple or orange markings.
SIZE: Carapace length to 9 cm.
DISTRIBUTION: Aleutian Islands to California. DEPTH: Subtidal in shallow rocky areas up to 183 m.
(Donaldson and Byersdorfer 2005)

Spiny lithode crab

Acantholithodes hispidus

DESCRIPTION: Medium-sized spiny crab. Carapace dorsal surface covered with small spines; rostrum sharp. Abdomen short, broad, calcified anteriorly and bearing many small spines. Carapace various shades of brown with opaque white areas, spines dark red-brown, orange, or white.
SIZE: Carapace length to 7.5 cm.
DISTRIBUTION: Aleutian Islands to California. DEPTH: Intertidal to 245 m.
(Donaldson and Byersdorfer 2005)

Rhinolithodes wosnessenskii

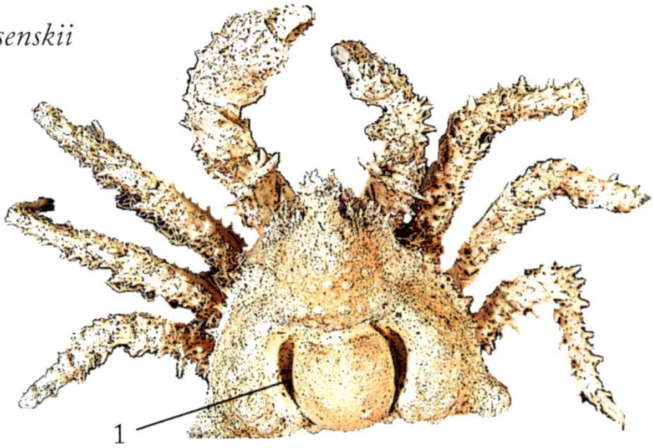

(Preserved)

Phyllolithodes papillosus

Acantholithodes hispidus

Soft crab

Hapalogaster grebnitzkii

DESCRIPTION: Small, soft, hairy crab. Carapace and claws covered with tufts of long brown hairs. Abdomen soft and loosely folded under body. Broad triangular rostrum (1); finger of right claw large, ventrally curved, and blue or black in color (2). Carapace red and orange; marginal spines with white tips.
SIZE: Carapace length to under 3 cm.
DISTRIBUTION: Bering Sea to British Columbia. DEPTH: Intertidal to 90 m.
(Donaldson and Byersdorfer 2005)

Hairy crab

Hapalogaster mertensii

DESCRIPTION: Small, soft, hairy crab. Body covered with tufts of long brown hairs. Abdomen soft and loosely folded under the body (1). Carapace brown and red with a few white spots. Claw fingers often reddish orange (2).
SIZE: Carapace length to 3.5 cm.
DISTRIBUTION: Atka, Alaska, to Washington. DEPTH: Low intertidal to 55 m.
(Donaldson and Byersdorfer 2005)

Butterfly crab

Cryptolithodes typicus

DESCRIPTION: Small cryptic crab. Carapace broad, completely covering legs; rostrum broader at the base than at end. Color variable, marked with red, black, white, and other colors.
SIZE: Carapace length to 4.1 cm.
DISTRIBUTION: Aleutian Islands to California. DEPTH: Low intertidal to 45 m on gravel slopes, rock, or shell debris.
(Donaldson and Byersdorfer 2005)

Hapalogaster grebnitzkii

Hapalogaster mertensii

Cryptolithodes typicus

Paralithodes crabs are distinguished by having the rostrum end in a single sharp-tipped projection and having the second abdominal segment covered by five distinct plates.

Red king crab

Paralithodes camtschaticus

DESCRIPTION: Very large king crab. Body and appendages covered with sharp thorns. Rostrum relatively long, with medial dorsal spine followed by a pair of spines. Middorsal plate of carapace usually has three pairs of prominent spines (1). Color red to reddish-purple dorsally, and white to cream ventrally.
SIZE: Carapace length to 23 cm.
DISTRIBUTION: Bering Sea to British Columbia. DEPTH: 3-400 m.
(Donaldson and Byersdorfer 2005, Clark 2006)

Blue king crab

Paralithodes platypus

DESCRIPTION: Large king crab. Rostrum short, ventrally slanted with a pair of dorsal spines; 2 pairs of prominent spines on the middorsal plate (1); 5 separate plates on second abdominal segment. Color blue or purple dorsally and white to cream ventrally.
SIZE: Carapace length to 16 cm.
DISTRIBUTION: Bering Sea to southeastern Alaska. DEPTH: 10-200 m.
(Kessler 1985)

Spiny king crab

Paralithodes brevipes

DESCRIPTION: Medium-sized king crab. Rostrum short, curved downward, and tip of lower process blunt; walking legs highly spinous; spines exceptionally long on internal edge of the chelipeds (1). Lateral edges of carapace bear large spines that form a distinct border. Scaphocerite has 4 spines; center of carapace has 2 pairs of spines. Color mottled reddish tones dorsally and white to cream ventrally.
SIZE: Carapace length to 9.5 cm.
DISTRIBUTION: Rare, Little Diomede Island and western Aleutians.
DEPTH: 0-50 m.
(Makarov 1962, Zaklan 2002)

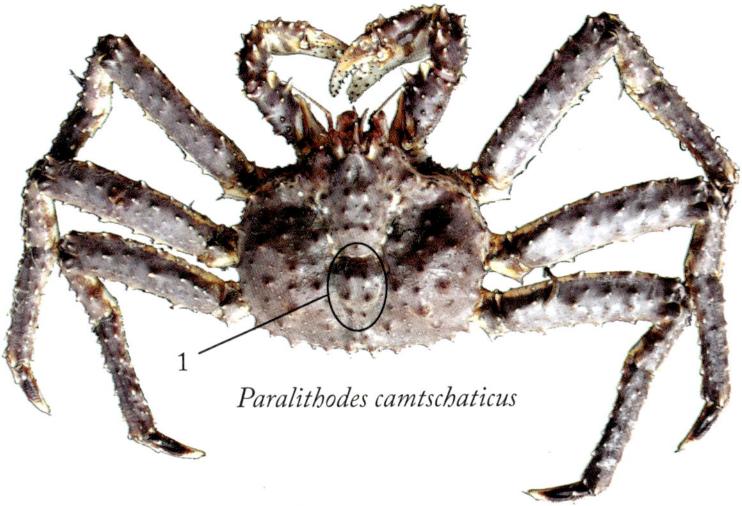

Paralithodes camtschaticus

Paralithodes platypus

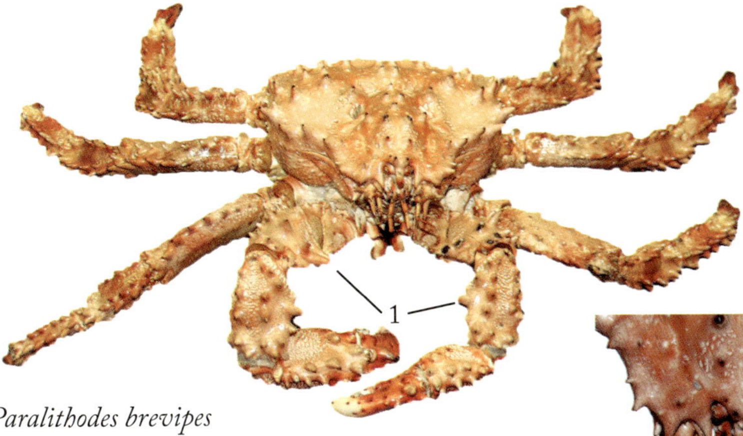

Paralithodes brevipes

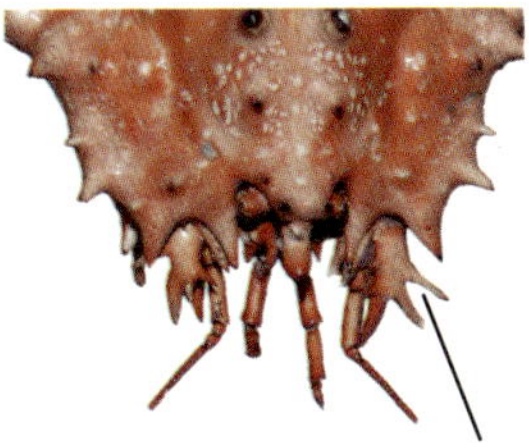

Lithodes crabs are distinguished by a rostrum that is divided into a paired tip and second abdominal segment plates that are completely or partly fused and are fewer than five in number.

Golden king crab

Lithodes aequispinus

DESCRIPTION: Very large king crab. Carapace covered in spines approximately equal in length. Spines of lateral margins are equal in length or slightly larger than those on dorsal surface. Rostrum has fairly broad base and is armed with 9 to 10 spines. Middorsal plate of carapace has 5 to 9 spines. Live specimens are golden on dorsal and ventral surfaces.
SIZE: Carapace length to 22 cm.
DISTRIBUTION: Bering Sea to British Columbia. DEPTH: 50-1,000+ m.
(Donaldson and Byersdorfer 2005)

Scarlet king crab

Lithodes couesi

DESCRIPTION: Large king crab. Carapace covered in spines of different lengths; at least 2 of the spines on the lateral margins are distinctly longer than the spines of the dorsal surface of the carapace (1). Rostrum long and slender, and armed with 7 spines. Middorsal plate is narrow relative to *L. aequispinus* and has 4 to 6 spines. Live specimens are brick red to pinkish on dorsal and ventral surfaces.
SIZE: Carapace length to 11.5 cm.
DISTRIBUTION: Bering Sea to Mexico. DEPTH: 200-2,000+ m.
(Donaldson and Byersdorfer 2005)

Scaled crab

Placetron wosnessenskii

DESCRIPTION: Large, scaled crab. Body and legs covered with scale-like plates; claws nearly equal in size. Abdomen thin, flat, mostly soft, and not completely covered by calcareous plates. Carapace red-brown medially and gray with dark-brown scales laterally.
SIZE: Carapace length to 6.2 cm.
DISTRIBUTION: Bering Sea to Washington. DEPTH: 5-110 m.
(Donaldson and Byersdorfer 2005)

Lithodes aequispinus

Lithodes couesi

Placetron wosnessenskii

Verrill's spider crab

Paralomis verrilli

DESCRIPTION: Large spider-like crab. Carapace slightly longer than wide; large spines on elevated areas and lateral margins. Walking legs compressed, edges with large flattened sawtooth spines (1) interspersed with scattered small spines. Pale pink or scarlet, carapace color often different from leg color.
SIZE: Carapace length to 10 cm.
DISTRIBUTION: Bering Sea to California. DEPTH: 800-2,000+ m.
(Donaldson and Byersdorfer 2005)

Spiny paralomis

Paralomis multispina

DESCRIPTION: Large spider-like crab. Carapace about as wide as it is long and thickly studded with blunt spines. Rostrum has a simple median spine with 2 basal spines; short conical spine under rostrum. Male abdomen has several rows of leathery plates. Ambulatory legs are long, slender, and thickly set with spines. Color uniform pale pink to scarlet.
SIZE: Carapace length to 8 cm.
DISTRIBUTION: Bering Sea to California. DEPTH: 800-2,000+ m.
(Donaldson and Byersdorfer 2005)

Spiny spider crab

Paralomis makarovi

DESCRIPTION: Large, spiny spider-like crab. Carapace pear-shaped; rounded posteriorly and longer than wide. Carapace and legs profusely covered with very long, slender spines. A uniform coverage of spines on the carapace and pereopods, with no dorsal spine more prominent than any other, easily distinguishes *P. makarovi* from other *Paralomis* crabs. Color light pink.
SIZE: Carapace length unknown. Carapace width to 10 cm.
DISTRIBUTION: Bering Sea and Aleutian Islands. DEPTH: Below 600 m.
(Clark 2006, Hall and Thatje 2009)

Paralomis verrilli

Paralomis multispina

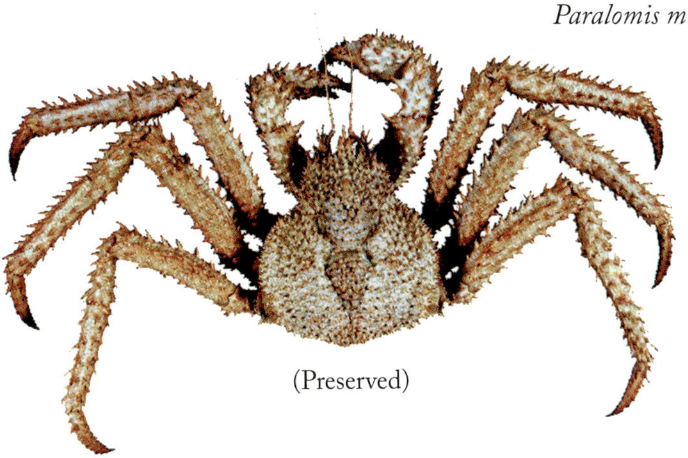

(Preserved)

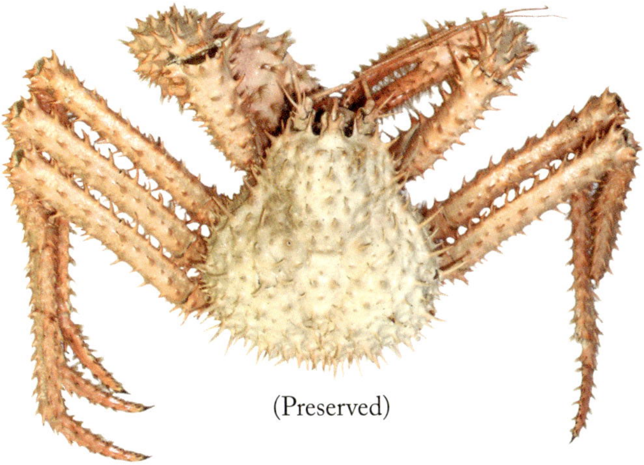

(Preserved)

Paralomis makarovi

Pinchbug

Munida quadrispina

DESCRIPTION: Small galatheid crab. Carapace with transverse folds or wrinkles. Rostrum long and slender, with a single basal spine located on each side. Four pairs of visible legs; body shrimp-like; first legs long. Claws long and narrow; eyes black; abdomen usually tucked under carapace. Color red to red-orange with white banding on legs.
SIZE: Carapace length to 3 cm.
DISTRIBUTION: Bering Sea to California. DEPTH: 20-600 m.
(Kessler 1985, Clark 2006)

Triangle Tanner crab

Chionoecetes angulatus

DESCRIPTION: Large spider crab. Carapace small, with a wide shallow valley bearing a central spinose ridge (1) between branchial regions. Single large spine at intersection of branchial ridges (2). Color orange to red-orange.
SIZE: Carapace width to 15 cm.
DISTRIBUTION: Bering Sea, adjacent to the Aleutian Islands, and Kamchatka Peninsula, to Oregon. DEPTH: 90-3,000 m.
(Kessler 1985, Jadamec et al. 1999)

Grooved Tanner crab

Chionoecetes tanneri

DESCRIPTION: Large spider crab. Carapace small, with a deep, narrow groove (1) between branchial regions. Carapace is very spinous. Two small, subequal spines are at the intersection of 2 dorsal branchial ridges (2). Color orange to orange-brown.
SIZE: Carapace width to 15 cm.
DISTRIBUTION: Bering Sea, primarily adjacent to the Aleutian Islands, to the Pacific Ocean off California-Mexico border. DEPTH: 53-1,900 m.
(Kessler 1985, Jadamec et al. 1999)

Munida quadrispina

Chionoecetes angulatus

Chionoecetes tanneri

Tanner crab

Chionoecetes bairdi

DESCRIPTION: Large spider crab. Carapace small, lacking central groove. Epistomal ridge (teeth) M-shaped (1); rostral spines triangular, sharply pointed, and tilted upward. Color pink to brown; eyes usually red.
SIZE: Carapace width to 19 cm.
DISTRIBUTION: Bering Sea to California. DEPTH: 5-800 m.
(Jadamec et al. 1999, Clark 2006)

Snow crab

Chionoecetes opilio

DESCRIPTION: Large spider crab. Carapace small, lacking central groove. Epistomal ridge (teeth) straight (1); rostral spines triangular, bluntly rounded, not tilted upward. Color tan to brown; eyes usually green or gray-green.
SIZE: Carapace width to 15 cm.
DISTRIBUTION: Bering Sea to western Gulf of Alaska. DEPTH: 5-800 m.
(Jadamec et al. 1999, Clark 2006)

Hybrid Tanner crab

Chionoecetes bairdi × *opilio*

DESCRIPTION: *Chionoecetes bairdi* and *C. opilio* hybridize where their ranges overlap, and offspring possess characteristics that are intermediate between the two species. The three most obvious intermediate characteristics of a hybrid crab are eye color (one red and one green); shape of the epistomal ridge (neither M-shaped nor horizontal); and the ratio of carapace length to carapace width (between 0.945 and 0.98).
DISTRIBUTION: Bering Sea. DEPTH: Subtidal to 450 m.
(Jadamec et al. 1999, Clark 2006)

Graceful decorator crab

Oregonia gracilis

DESCRIPTION: Small spider crab. Carapace small, 5 pairs of legs visible. Body triangular, profusely decorated usually with pieces of kelp, seaweeds, hydroids, sponges, or bottom debris. Rostral spines long, thin, joined for part of their length, and separated by a narrow slit. Color gray or light brown.
SIZE: Carapace length to 5 cm.
DISTRIBUTION: Chukchi Sea to California. DEPTH: 0-100 m.
(Hart 1982, Kessler 1985)

Chionoecetes bairdi

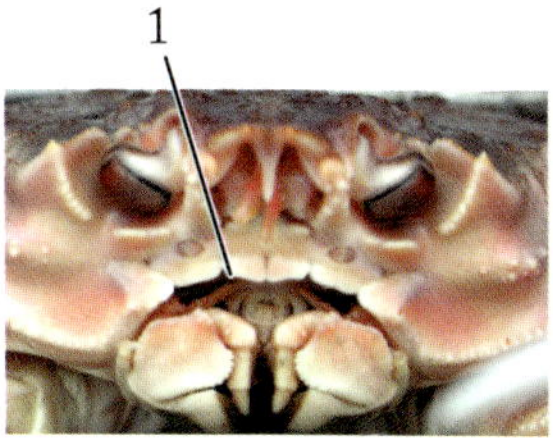

Chionoecetes opilio

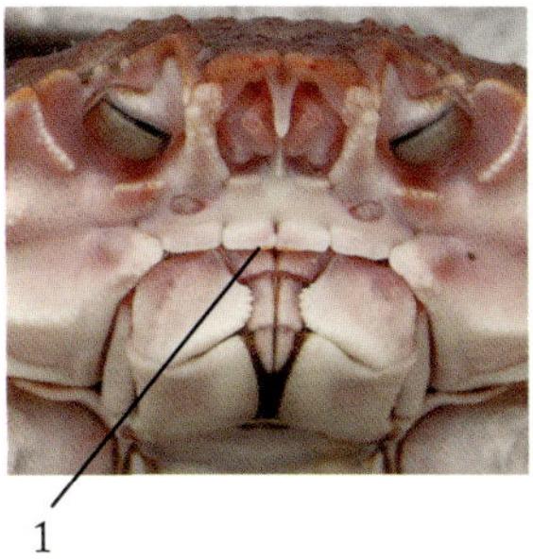

Chionoecetes bairdi × *Chionoecetes opilio*

Oregonia gracilis

Completely decorated

Lightly decorated

Pacific lyre crab

Hyas lyratus

DESCRIPTION: Medium-sized spider crab. Carapace lyre-shaped with tubercles and a winglike lateral expansion behind each eye. Forward lateral projection sharp (1), with a prominent posterior spine; edges of expansions are toothed. Rostrum divided into 2 flattened, blunt-tipped horns. Five pairs of visible legs. Chelipeds (2) of mature males are stout and longer than walking legs. Legs are long, slender, and cylindrical; leg color white, orange, and brown. Carapace color brown, marked with white and orange.

SIZE: Carapace length to 8 cm.

DISTRIBUTION: Aleutian Islands to Washington. DEPTH: 10-640 m.

(Hart 1982, Jensen 1995)

Arctic lyre crab

Hyas coarctatus

DESCRIPTION: Medium-sized spider crab. Carapace lyre-shaped with a smooth posterior. Forward lateral projection of carapace rounded (1), no posterior spine. Five pairs of visible legs. Chelipeds (2) of mature males are stout and longer than walking legs. Color maroon marked with white and orange.

SIZE: Carapace length to 8 cm.

DISTRIBUTION: Arctic Ocean to northern Gulf of Alaska. DEPTH: 10-400 m.

(Kessler 1985, Clark 2006)

Graceful kelp crab

Pugettia gracilis

DESCRIPTION: Small spider crab. Carapace with tubercles and a prominent posterior projection (1). Dorsal surface of merus of first leg has irregularly toothed ridge (2). Color reddish and olive tones, often with blue on claws.

SIZE: Carapace length to 6 cm.

DISTRIBUTION: Aleutian Islands to California. DEPTH: 0-200 m.

(Kessler 1985, Clark 2006)

Hyas lyratus

Hyas coarctatus

Pugettia gracilis

Longhorn decorator crab

Chorilia longipes

DESCRIPTION: Small spider crab. Carapace pear-shaped, with spaced, short tubercles or spines. Rostral spines relatively long, stout, sharp, and divergent. Walking legs long, smooth and cylindrical; first longer than cheliped. Carapace white with orange spines; rostrum orange. Chelipeds orange-red; finger scarlet with white band and white tips (1). Walking legs red with white bands.
SIZE: Carapace length to 6 cm.
DISTRIBUTION: Bering Sea to Mexico. DEPTH: 10-1,500 m.
(Hart 1982, Clark 2006)

Hair crab

Erimacrus isenbeckii

DESCRIPTION: Large hairy crab. Body about as long as wide; margin of carapace posterior to the eye has 7 lateral spines. Long, shaggy hair covers the underside of the legs and abdomen. Color reddish brown to brown.
SIZE: Carapace length to 13 cm.
DISTRIBUTION: Bering Sea to western Gulf of Alaska. DEPTH: 2-350 m.
(Kessler 1985, Clark 2006)

Helmet crab

Telmessus cheiragonus

DESCRIPTION: Large hairy crab. Chelipeds short and spiny with claw tips black (1); body slightly wider than long. Margin of carapace posterior to the eye has 6 lateral spines. Color yellowish or golden-brown with red speckles.
SIZE: Carapace length to 10 cm.
DISTRIBUTION: Bering Sea to California. DEPTH: 0-300 m.
(Hart 1982, Kessler 1985)

Chorilia longipes

Erimacrus isenbeckii

Female
abdominal flap

Male
abdominal flap

Telmessus cheiragonus

Dungeness crab

Cancer magister

DESCRIPTION: Large ovate crab. Body oval with carapace almost twice as wide as long; widest at the last (or 10th) spine along carapace (1). Five pairs of visible legs. Carapace color light brown or olive and purple. Ridges on claws rough, tips light-colored (2).
SIZE: Carapace width to 27 cm.
DISTRIBUTION: Southeastern Bering Sea to Mexico. DEPTH: 0-460 m.
(Kessler 1985, Clark 2006)

Pygmy rock crab

Cancer oregonensis

DESCRIPTION: Small stout crab. Carapace smooth or very rough, margins rounded, not clearly defined into anterior and posterior like other species of *Cancer.* Claw tips black (1); chelipeds white with orange; fingers black. At least 3 variants may be found. The common form has marginal spines that are closely set and curved (2). Carapace red-brown with gray granules and white patches.
SIZE: Carapace width to 5 cm.
DISTRIBUTION: Bering Sea to California. DEPTH: 0-440 m.
(Hart 1982)

Gaper pea crab

Pinnixa littoralis

DESCRIPTION: Males have a firm carapace that is about twice as wide as long, and chelipeds with a gap between fingers. Females have an oval-shaped soft carapace. Both sexes with outer margin of eye orbits pointed laterally. Merus of third walking legs slightly less than twice as long as wide; dactyls of all walking legs strongly curved and sharply pointed. Color of both sexes white; males often with darker markings.
SIZE: Carapace width—male 1.8 cm, female 2.7 cm.
DISTRIBUTION: Prince William Sound, Alaska, to California. Found in the mantle cavity of the fat gaper clam *Tresus capax*. DEPTH: 0-20 m.
(Hart 1982)

Cancer magister

Cancer oregonensis

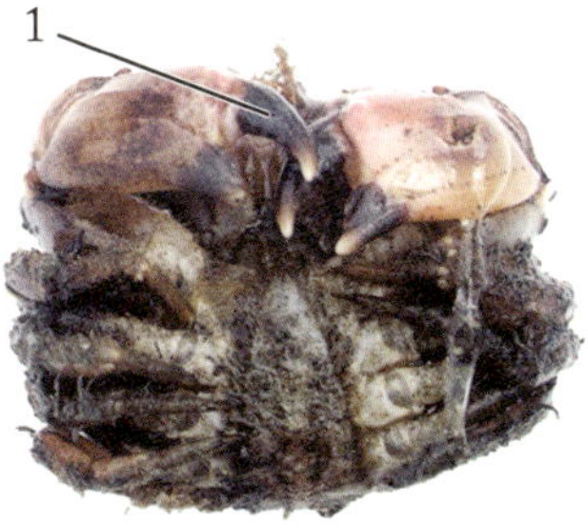

Pinnixa littoralis

Female

Male

HERMIT CRABS

Hermit crabs do not have a very hard shell, nor are they true crabs. They use discarded snail shells to protect their soft abdomens. As they grow in size, they must find a larger shell for protection. Worldwide there are about 500 species of hermit crabs, most of which live on the ocean floor. Many hermit crabs live on land, however, and the females must return to the sea to breed.

HERMIT CRAB ANATOMY

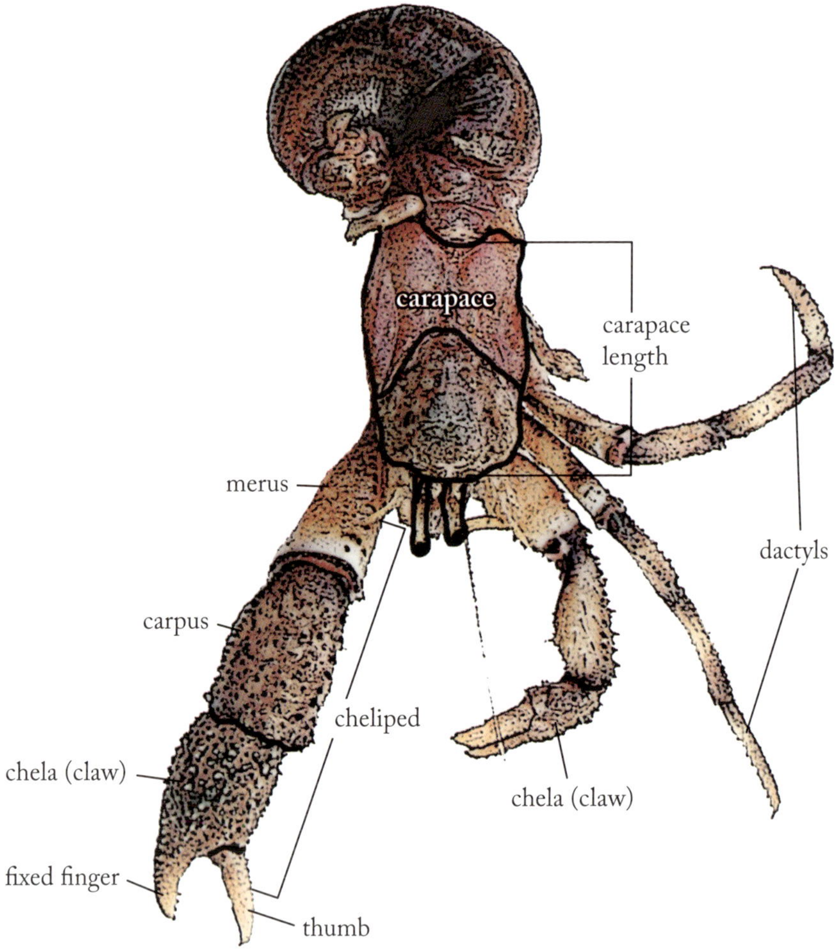

Splendid hermit

Labidochirus splendescens

DESCRIPTION: Medium-sized hermit crab. Rear of carapace calcified, eyestalks short. Chelipeds unequal, right considerably larger than left, elongate and slender. Iridescent maroon or lavender color. Often found in moonsnail shells encrusted with a brown, velvety hydroid growth.
SIZE: Carapace length to 3 cm.
DISTRIBUTION: Arctic Ocean, Chukchi and Bering seas, Gulf of Alaska to Washington.
DEPTH: 3-450 m.
(McLaughlin 1974, Kessler 1985, Clark 2006)

Labidochirus splendescens

Widehand hermit

Elassochirus tenuimanus

DESCRIPTION: Medium-sized hermit crab. Right claw wide, with a prominent lateral projection; carpus narrower than claw. Color reddish-brown to tan, merus of legs often bright blue or purple (1).
SIZE: Carapace length to 4 cm.
DISTRIBUTION: Bering Sea, Aleutian Islands to Washington.
DEPTH: 1-246 m.
(McLaughlin 1974, Kessler 1985, Clark 2006)

Elassochirus tenuimanus

Purple hermit

Elassochirus cavimanus

DESCRIPTION: Medium-sized hermit crab. Right claw narrow, compressed (1); carpus wider than claw, with winglike projections (2). Right carpus with one or more tiny irregular spines on dorsal surface. Color of carpus and merus of claws purple.
SIZE: Carapace length to 4 cm.
DISTRIBUTION: Bering Sea, Aleutian Islands, Gulf of Alaska to Washington.
DEPTH: 25-260 m.
(McLaughlin 1974, Kessler 1985, Clark 2006)

Pacific red hermit

Elassochirus gilli

DESCRIPTION: Medium-sized hermit crab. Similar to *Elassochirus cavimanus* but the first right carpus lacks tiny spines on the dorsal surface. Nearly uniformly colored a bright red-orange.
SIZE: Carapace length to 4 cm.
DISTRIBUTION: Bering Sea, Aleutian Islands, Gulf of Alaska to Washington.
DEPTH: 0-200 m.
(McLaughlin 1974, Kessler 1985, Clark 2006)

Aleutian hermit

Pagurus aleuticus

DESCRIPTION: Large hermit crab. Both claws flattened, with raised margins. Right cheliped stout, and shorter than walking legs; carpus and hand covered with spines and granules, with serrated outer margins. Left cheliped slender. Dactyls long, stout and twisted; orange-colored with a red stripe on the outer face and a dark red dorsal groove. Uniform reddish-brown to orange in color.
SIZE: Carapace length to 5 cm.
DISTRIBUTION: Bering Sea, Aleutian Islands, Kodiak Island and Gulf of Alaska.
DEPTH: 10-100+ m.
(McLaughlin 1974, Hart 1982, Clark 2006)

Elassochirus cavimanus

Elassochirus gilli

Pagurus aleuticus

Alaskan hermit

Pagurus ochotensis

DESCRIPTION: Large hermit crab. Both claws flattened, with raised margins. Spines on claws short and blunt; lower side of left claw armed with small spines. Prominent red stripe under thumb (1); merus with iridescent blue or purple areas (2); eyes yellow green.
SIZE: Carapace length to 5 cm.
DISTRIBUTION: Bering Sea, Pribilof Islands, Gulf of Alaska, southeastern Alaska to California. DEPTH: 1-400 m.
(McLaughlin 1974, Kessler 1985, Clark 2006)

OTHER HERMIT CRABS WITH DORSALLY FLATTENED CLAWS AND RAISED MARGINS. The armed hermit *Pagurus armatus* resembles the Alaskan hermit but lacks the red stripe under the thumb and the spines on ventral side of left claw. It has black eyes and longer claw spines than the Alaskan hermit.

Longfinger hermit

Pagurus rathbuni

DESCRIPTION: Medium-sized hermit crab. Both claws convex and gently rounded. Long ventrally curved thumbs and fingers. Males positively identified by a patch of hairlike setae set in a depression on upper left corner of carpus (1) behind the right claw; females have depression but lack setae. Color cream to tan with light orange-brown banding.
SIZE: Carapace length to 4 cm.
DISTRIBUTION: Chukchi Sea, Bering Sea, Pribilof Islands. DEPTH: 30-210 m.
(McLaughlin 1974, Kessler 1985, Clark 2006)

"Dirty harry" hermit

Pagurus capillatus

DESCRIPTION: Medium-sized hermit crab. Both claws convex and gently rounded. Claws not black-tipped and are completely covered with dense hairlike setae that obscures spination; setae are usually saturated with mud. Antennae without bands, merus without white band; a row of fine spines on ventral margin of second and third dactyls. Color brown or orange tones.
SIZE: Carapace length to 4 cm.
DISTRIBUTION: Arctic Ocean, Chukchi Sea, Bering Sea, Aleutian Islands to Gulf of Alaska. DEPTH: 4-236 m.
(McLaughlin 1974, Kessler 1985, Stevens 2002, Clark 2006)

Pagurus ochotensis

Pagurus rathbuni

Pagurus capillatus

OTHER HERMIT CRABS WITH GENTLY ROUNDED CLAWS. The setose hermit *Pagurus setosus* resembles the "dirty harry" hermit but the setae are not dense enough to obscure claw spination or the large spines on the ventral margin of second and third dactyls. The bluespined hermit *P. kennerlyi* has black-tipped claws with coarse blue-tinted spines; hairlike setae are not dense enough to obscure claw spination.

Fuzzy hermit

Pagurus trigonocheirus

DESCRIPTION: Large hermit crab. Claws often covered with fine hairlike setae or golden-brown fuzz. First right carpus about 1.5 times longer than wide. Color cream to tan; merus end with dark coloration (1).
SIZE: Carapace length to 5 cm.
DISTRIBUTION: Arctic Ocean, Chukchi Sea, Bering Sea, Aleutian Islands. Pribilof Islands to Gulf of Alaska. DEPTH: 20-182 m.
(McLaughlin 1974, Kessler 1985, Clark 2006)

Whiteknee hermit

Pagurus dalli

DESCRIPTION: Small hermit crab. Left claw with a sharp median crest; no prominent ridges on right claw; diameter of eyes smaller than stalks; claws not covered with hairlike setae. Dorsal carapace has a network of white, tan, mahogany, and red colors. There is a distinct white band (1) on the distal parts of the merus of the chelipeds, which is the source of the animal's common name.
SIZE: Carapace length to 3 cm.
DISTRIBUTION: Bering Sea, Aleutian Islands to Oregon. DEPTH: Low intertidal to 183 m.
(McLaughlin 1974, Kessler 1985)

Knobbyhand hermit

Pagurus confragosus

DESCRIPTION: Medium-sized hermit crab. Both claws with prominent ridges or crests. Right claw less than twice as long as wide, with a prominent knob (1); ridge on right claw extends into finger as a flattened plateau. Color mottled red or brown.
SIZE: Carapace length to 4 cm.
DISTRIBUTION: Bering Sea, Gulf of Alaska south to Oregon. DEPTH: 30-435 m.
(McLaughlin 1974, Kessler 1985)

OTHER HERMIT CRABS WITH PROMINENT RIDGES ON BOTH CLAWS. The Pribilof hermit *Pagurus undosus* is distinguished by prominent sculpturing on the right claw that looks like a gently rounded ridge rather than a triangular plateau. The longhand hermit *P. tanneri* is distinguished in that the ridge of the right claw extends onto finger as a slender crest and the right claw is elongate (more than twice as long as wide). The hornyhand hermit *P. cornutus* is distinguished by a prominent horn on the right claw.

Pagurus trigonocheirus

Pagurus dalli

Pagurus confragosus

BRYOZOANS

Bryozoans are colonial animals; most are attached to substrates, while others are erect and branched. Bryozoans can also resemble hydroids, so careful observation is necessary to separate the two groups. Most colonies are white to tan, but a few are pigmented. In many species of bryozoans, not all individuals in a colony look identical. Without a hand lens or microscope to identify anatomical features, few bryozoans can be identified to species. Other bryozoans similar to those illustrated may occur in this region.

Jointed bryozoan

Microporina borealis

DESCRIPTION: Medium-sized, branching bryozoan; jointed colonies resemble strings of sausages; occurs in clumps. Color orange to brown.
SIZE: Height to 15 cm.
DISTRIBUTION: Circumboreal, south to Washington. DEPTH: 5-400 m.
(Clark 2006)

Fan bryozoan

Dendrobeania sp.

DESCRIPTION: Medium-sized, flattened bryozoan; fan-like branching colonies; occurs in clumps. Color light brown, lighter at branch tips.
SIZE: Height to 8 cm.
DISTRIBUTION: Bering Sea to Washington. DEPTH: 1-140 m.
(Clark 2006)

Spiral bryozoan

Bugula californica

DESCRIPTION: Medium-sized delicate bryozoan; bottlebrush-shaped colonies grow spirally and often occur in clumps. Color pale orange, yellow-tan, or brown.
SIZE: Height to 8 cm.
DISTRIBUTION: Aleutian Islands to Mexico. DEPTH: 1-130 m.
(Clark 2006)

Microporina borealis

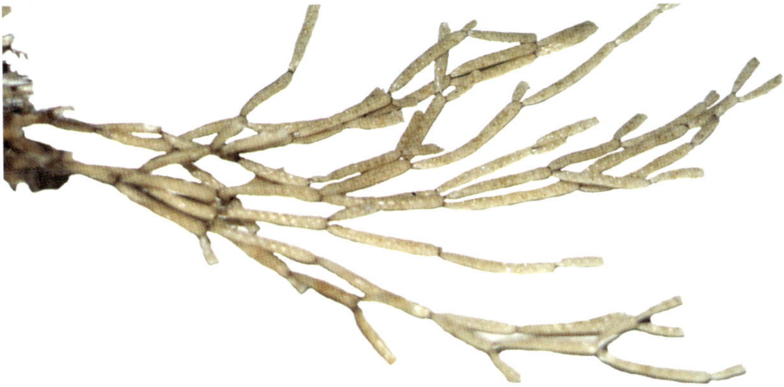

Dendrobeania sp.

Bugula californica

Smooth leather bryozoan

Alcyonidium pedunculatum

DESCRIPTION: Large, upright bryozoan; rubbery, sheetlike colonies are smooth to the touch. Color tan.
SIZE: Height to 30 cm.
DISTRIBUTION: Aleutian Islands to British Columbia. DEPTH: 20-60 m.
(Clark 2006)

Medusa bryozoan

Alcyonidium sp.

DESCRIPTION: Large rubbery bryozoan; bush-like colonies with numerous multiple-forked, blunt-ended branches. Color yellowish-tan.
SIZE: Height to 12 cm.
DISTRIBUTION: Eastern Aleutian Islands to Gulf of Alaska. DEPTH: 20-120 m.
(Clark 2006)

Spiny leather bryozoan

Flustrellidra corniculata

DESCRIPTION: Medium-sized, leathery bryozoan; spine-covered colonies are fuzzy-looking with a soft, leather-like texture. Color tan to brown.
SIZE: Height to 10 cm.
DISTRIBUTION: Aleutian Islands to California. DEPTH: 1-75 m.
(Clark 2006)

Alcyonidium pedunculatum

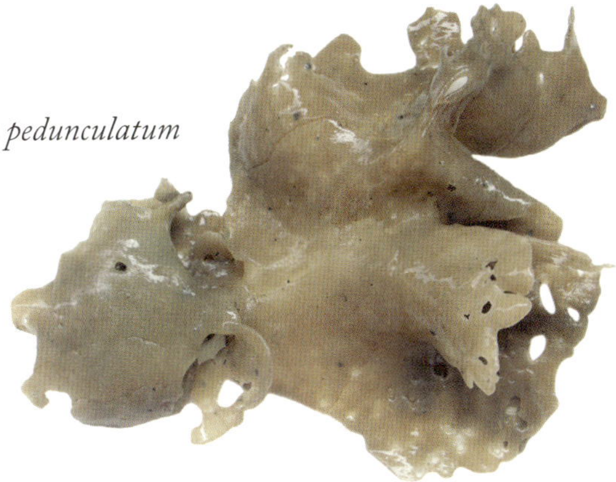

Alcyonidium sp.

Flustrellidra corniculata

FEATHER STAR

Feather star

Florometra sp.

DESCRIPTION: Medium-sized, unstalked feather star (crinoid). Crown with 10 long, slender, feather-like arms; cirri numerous, long, and slender. Color uniform shades of yellow or brown.

SIZE: Height to 23 cm.

DISTRIBUTION: Aleutian Islands to Gulf of Alaska. DEPTH: 0-2,000 m.

(Clark 2006)

Florometra sp.

SEA STARS

EXTERNAL FEATURES OF SEA STARS

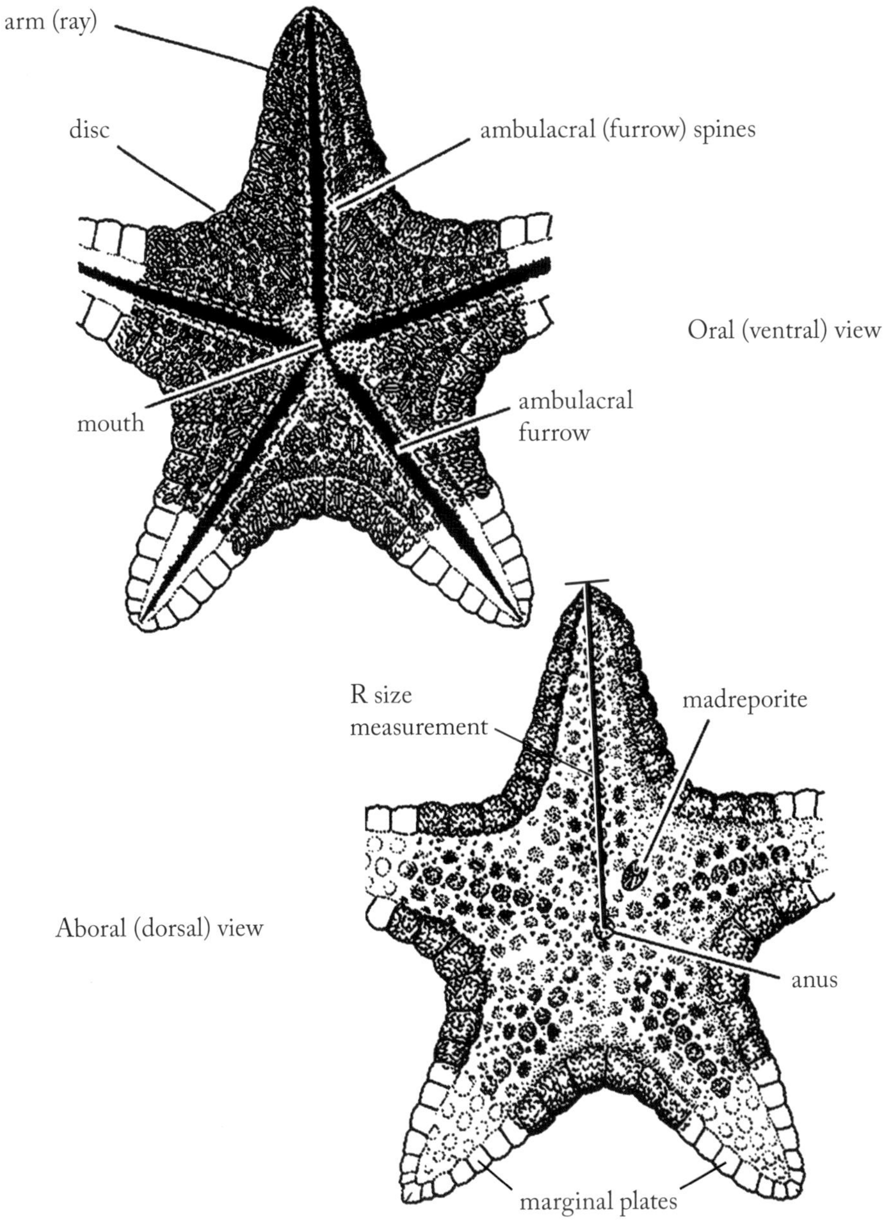

Adapted from Spencer and Wright 1966

Arctic sand star

Leptychaster arcticus

DESCRIPTION: Small star; disc broad, arms long, tapering. Aboral paxillae (spines) minutely spinose, densely packed. Supramarginal plates very small (1) in relation to inframarginal plates (2), which are visible when viewed from above. Color orange aborally, white orally, arms sometimes white.
SIZE: R to 5 cm.
DISTRIBUTION: Circumboreal, Arctic Ocean, Bering Sea south to Oregon.
DEPTH: 40-1,260 m.
(Lambert 2000, Clark 2006)

Common mud star

Ctenodiscus crispatus

DESCRIPTION: Small star; arms outlined by prominent marginal plates, each with a single upturned thorn (1); lower plates not visible from above; no prominent spines. Ball-like plate on end of each ray; cone in center of disc (2). Color uniformly yellow-tan.
SIZE: R to 5.5 cm.
DISTRIBUTION: Circumboreal, south to Panama. DEPTH: 10-1,900 m.
(Kessler 1985, Clark 2006)

Fragile star

Cheiraster dawsoni

DESCRIPTION: Medium-sized star; long slender arms, tapering to fine points that are soft and easily broken; disc small. Arms outlined by prominent marginal plates; aboral surface with prominent fine spines (1). Color red-orange to orange, often with white mottling aborally, white orally.
SIZE: R to 20 cm.
DISTRIBUTION: Bering Sea to California. DEPTH: 70-400 m.
(Kessler 1985, Clark 2006)

Leptychaster arcticus

Ctenodiscus crispatus

Cheiraster dawsoni

Orange cookie star

Ceramaster patagonicus

DESCRIPTION: Broad, flat pentagonal star with arms tapering to slender tips; interradial margins usually flat (1). Aboral surface often swollen and soft to the touch. Dorsal polygonal plates (2) with 4-12 central and 12-15 marginal granules. Color orange aborally, whitish orally.
SIZE: R to 12 cm.
DISTRIBUTION: Bering Sea, southeastern Alaska to British Columbia.
DEPTH: 10-250 m.
(Lambert 2000, Clark 2006)

Arctic cookie star

Ceramaster arcticus

DESCRIPTION: Broad, flat, pentagonal star, stiff and firm to the touch; oral surface granular between arms. Dorsal polygonal plates (1) larger and less numerous than on red or orange cookie stars; 2 or 3 groove spines on each plate along the ventral furrow; marginal plates large and prominent. Color reddish, pink, lavender, or pale orange, with darker markings aborally, paler orally.
SIZE: R to 6 cm.
DISTRIBUTION: Bering Sea, Aleutian Islands, and Gulf of Alaska.
DEPTH: 0-186 m.
(Kessler 1985, Lambert 2000, Clark 2006)

Stellate cookie star

Ceramaster stellatus

DESCRIPTION: Broad, flat star with evenly tapering arms; interradial margins rounded (1). Dorsal polygonal plates distinctly hexagonal (2); with 10 to 12 marginal plates and 4 to 7 central granules. Marginal plates have small pedicellariae. Plates along ventral furrow with 4 robust spines. Color orange aborally; white orally.
SIZE: R to 8 cm.
DISTRIBUTION: Okhotsk and Bering seas, Aleutian Islands. DEPTH: Below 400 m.
(Clark 2006)

Ceramaster patagonicus

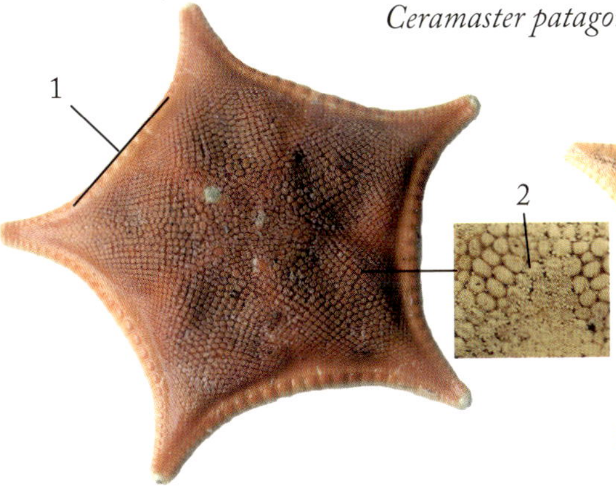

Ceramaster arcticus

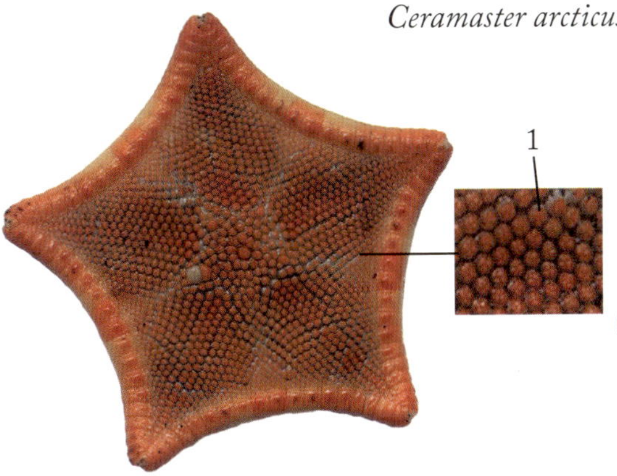

Ceramaster stellatus

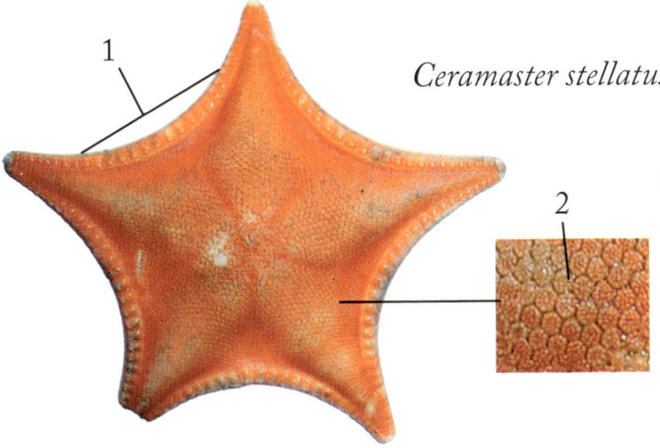

Red cookie star

Ceramaster japonicus

DESCRIPTION: Broad, flat, pentagonal star; dorsal polygonal plates and marginal plates bearing pedicellariae. Oral surface smooth between arms (1). Color red to red-orange aborally, cream-white orally.
SIZE: R to 10 cm.
DISTRIBUTION: Bering Sea, Gulf of Alaska, North Pacific Ocean to Oregon.
DEPTH: 200-1,400 m.
(Clark 2006)

Grainy white cookie star

Cladaster validus

DESCRIPTION: Broad, flat, stellate star with evenly tapering arms; aboral and oral marginal plates bearing spherical granules. Color uniform, white to pale pink.
SIZE: R to 3.5 cm.
DISTRIBUTION: Commander, Aleutian, and Shumagin islands to British Columbia. DEPTH: 200-500 m.
(Lambert 2000, Clark 2006)

Kurile spiny star

Hippasteria kurilensis

DESCRIPTION: Broad star with arms tapering to slender tips; aboral and marginal plates with sharp, thornlike granules around the margin; and bearing slender, tapering spines (2 on marginal plates; sparse on surface) (1). Plates along ventral furrow with 2 spines. Color orange with white markings aborally; white orally.
SIZE: R to 10 cm.
DISTRIBUTION: Kurile and Aleutian islands. DEPTH: 100-400 m.
(Clark 2006)

Ceramaster japonicus

Cladaster validus

Hippasteria kurilensis

Pale spiny star
Hippasteria sp.

DESCRIPTION: Broad star with broad arms tapering to slender tips. Marginal plates with 2 to 4 pointed spines, edged with flattened, pointed granules; aboral surface profusely covered with short, slender spines. Plates along ventral furrow with a single spine. Color beige with a few orange markings aborally; beige orally.
SIZE: R to 8 cm.
DISTRIBUTION: Aleutian Islands. DEPTH: 300-500 m.
(Clark 2006)

Northern scarlet star
Pseudarchaster parelii

DESCRIPTION: Large flat star; arms slender. Disc small; lower marginal plates near disc have minute thornlike projection. Granules on aboral surface arranged in polygons that are more crowded and less distinct toward tip. Mouth large, obvious; large unpaired spine at inner junction of each pair of mouthparts. Color uniform orange-red to burgundy aborally, tan to cream orally.
SIZE: R to 10 cm.
DISTRIBUTION: Circumboreal, Bering Sea, Aleutian Islands, Kodiak Island, Gulf of Alaska south to Oregon. DEPTH: 40-700 m.
(Kessler 1985, Lambert 2000, Clark 2006)

Alaskan scarlet star
Pseudarchaster parelii alascensis

DESCRIPTION: Medium-sized flat star; arms very long, slender. Marginal plates about ¼ as long as wide, lower marginal plates near disc with short, upward-directed thorns. Color uniform purple-red aborally, tan to cream orally.
SIZE: R to 8 cm.
DISTRIBUTION: Gulf of Alaska to northern California. DEPTH: 90-2,000 m.
(Clark 2006)

Hippasteria sp.

Pseudarchaster parelii

Pseudarchaster parelii alascensis

Leather star

Dermasterias imbricata

DESCRIPTION: Very large star; the only 5-armed star that has a soft, smooth to slippery aboral surface. Tough leather-like skin smells like garlic. Color mottled gray-green and reddish brown.
SIZE: R to 20 cm.
DISTRIBUTION: Gulf of Alaska to California. DEPTH: 0-200 m.
(Gotshall and Laurent 1979, Kessler 1985, Clark 2006)

Striped sun star

Solaster stimpsoni

DESCRIPTION: Very large sun star; usually has 10 long and slender arms (range 9 to 11). Color pattern distinctive, center of disc, top of arms, and oral surface purple, red, or violet; sides of arms red, orange, or yellow.
SIZE: R to 20 cm.
DISTRIBUTION: Bering Sea, south of the Alaska Peninsula, Kodiak Island to Oregon. DEPTH: 0-610 m.
(Kessler 1985, Lambert 2000, Clark 2006)

Remarkable sun star

Solaster sp.

DESCRIPTION: Large sun star with 8 arms; disc broad, arms rather short, broad. Color white or pale tan with bright red star in center of disc in Aleutian Island specimens; uniformly pale pink in Gulf of Alaska stars.
SIZE: R to 13 cm.
DISTRIBUTION: Bering Sea, Aleutian Islands, Gulf of Alaska. DEPTH: 70-400 m.
(Kessler 1985, Clark 2006)

Dermasterias imbricata

Solaster stimpsoni

Solaster sp.

Beautiful sun star

Solaster sp.

DESCRIPTION: Very large, 9 to 11 armed sun star; disc broad; arms moderately long, tapering to slender tips. Aboral paxillae small, spaced, often coalescing. Color highly variable, may be uniformly white, reddish-brown, or purple, often with rose markings on disc and arm tips; a common variant has a yellow-orange disc and purple arms.
SIZE: R to 20 cm.
DISTRIBUTION: Central Aleutian Islands. DEPTH: 80-370 m.
(Clark 2006)

Serpent sun star

Solaster sp.

DESCRIPTION: Small, eight-armed star; disc small, arms long, slender; aboral paxillae small, round, close-set. Color always uniformly white aborally and orally.
SIZE: R to 7.5 cm.
DISTRIBUTION: Western Aleutian Islands. DEPTH: 120-330 m.
(Clark 2006)

Rose star

Crossaster papposus

DESCRIPTION: Small, 10 or 11 armed star. Spinelets in the form of long brush-like tufts scattered irregularly over aboral surface. Color variable, usually in shades of rose, red, or pink aborally, white orally; markings are primarily in concentric bands of bright colors on the aboral surface.
SIZE: R to 8 cm.
DISTRIBUTION: Circumboreal, Arctic Ocean to northern California.
DEPTH: 1-1,200 m.
(Kessler 1985, Lambert 2000)

OTHER SPECIES OF *SOLASTER* AND *CROSSASTER*. The northern sun star *Solaster endeca* has 8-13 short inflated rays and numerous spinelets in short erect tufts. The morning sun star *S. dawsoni* is very large, with 12-14 rays; marginal spinelets not prominent, and ventral space between the rays is narrow with webbed spinelets. The evening sun star *S. paxillatus* has 9 rays, with prominent, well-spaced marginal spinelets. The grooved sun star *Crossaster borealis* has 11-16 rays with a distinct narrow groove between each ray.

Solaster sp.

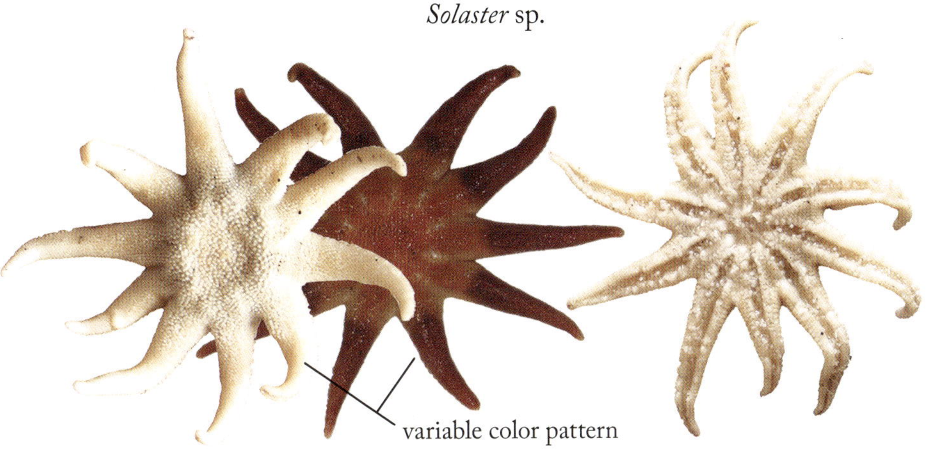

Solaster sp.

Crossaster papposus

Crested star

Lophaster furcilliger

DESCRIPTION: Medium-sized star; disc broad and relatively rigid; arms short, broad at base, tapering to rather blunt tips. Skeletal mesh open; aboral paxillae tall, fairly robust, spinose. Color pale yellow to cream, usually with dark pink ray radiating out from disc center to ray tips aborally, cream or pale yellowish orally.
SIZE: R to 6 cm.
DISTRIBUTION: Southern Bering Sea, south of the Alaska Peninsula to California. DEPTH: 21-2,000 m.
(Lambert 2000, Clark 2006)

Tessellated slime star

Pteraster tesselatus

DESCRIPTION: Large inflated stellate star; hard, soft, or sponge-like to the touch. Arms thick, short to long, and without spines; upturned at tips to reveal the ambulacral furrow. Prominent central osculum (1). No prominent marginal plates; actinolateral membrane very narrow (2). Body covered with a membrane; secretes lots of mucus when disturbed; tube feet in 2 rows. Color cream, buff, yellow, or light brown with orange, dark-brown, or bluish markings aborally; cream, yellow, or buff orally.
SIZE: R to 10 cm.
DISTRIBUTION: Bering Sea south to California. DEPTH: 9-450 m.
(Kessler 1985, Lambert 2000, Clark 2006)

Northern slime star

Pteraster sp.

DESCRIPTION: Large inflated, pentagonal star. Aboral surface thick, firm; actinolateral membrane very narrow (1). Produces copious amounts of mucus when disturbed. Color white or very pale orange, with orange or purple markings aborally; white orally.
SIZE: R to 7 cm.
DISTRIBUTION: Aleutian Islands. DEPTH: 80-400 m.
(Clark 2006)

Lophaster furcilliger

Pteraster tesselatus

Pteraster sp.

Obscure cushion star

Pteraster obscurus

DESCRIPTION: Large inflated star with 6 or 7 arms. Aboral surface thick, firm; actinolateral membrane wide, defining ray regions. Color buff or grayish with dark brown with purple-gray markings aborally, cream orally.
SIZE: R to 6 cm.
DISTRIBUTION: Circumboreal, Bering Sea and eastern Aleutian Islands.
DEPTH: 25-175 m.
(Kessler 1985, Clark 2006)

Eight-rayed cushion star

Pteraster octaster

DESCRIPTION: Large, inflated star with 8 or 9 arms; aboral surface firm, thick, usually with numerous scattered, small, low conical projections; actinolateral membrane wide, defining ray region. Color yellowish or yellow-orange with dark brown or purplish marking aborally, yellow orally.
SIZE: R to 6 cm.
DISTRIBUTION: Okhotsk Sea, Kamchatka, and the Commander Islands; Pribilof Islands and western Bering Sea. DEPTH: 20-135 m.
(Smirnov 1982, Clark 2006)

Wrinkled cushion star

Pteraster militaris

DESCRIPTION: Medium-sized, inflated, stellate star with a large pore in the center of the disc. Arms long and stubby, aboral surface soft, fleshy, and often wrinkled. Actinolateral membrane wide, defining ray margins. Color uniform, cream, yellow, orange, or pink aborally; tube feet pale.
SIZE: R to 6 cm.
DISTRIBUTION: Circumboreal, Bering Sea to Oregon. DEPTH: 10-1,100 m.
(Lambert 2000, Clark 2006)

Pincushion star

Diplopteraster multipes

DESCRIPTION: Large, spiny cushion-like star; 5 short stubby arms. Aboral surface covered with numerous protruding white spines. Tube feet in 4 rows. Secretes large amount of mucus when disturbed. Color brown to orange brown, margins cream, oral surface pale orange or cream, tube feet orange.
SIZE: R to 12 cm.
DISTRIBUTION: Circumboreal, Arctic Ocean to southern California.
DEPTH: 90-1,200 m.
(Kessler 1985, Clark 2006)

Pteraster obscurus

Pteraster octaster

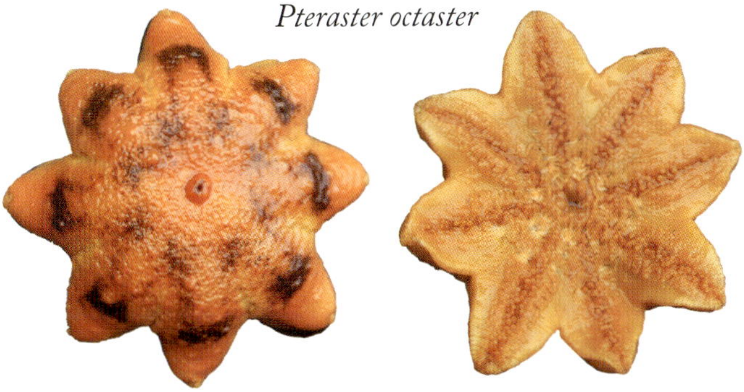

Pteraster militaris

Diplopteraster multipes

Blood star

Henricia cf. *leviuscula*

DESCRIPTION: Medium-sized star; disc small, long tapering arms almost round in cross section. Smooth arms lack spines and pedicellariae. Usually has 5 arms, but occasionally 4 or 6. Skeletal mesh fine; aboral plates variable in form, small, close-set, topped with rounded granules; three rows of plates along the lower side of each arm. Color brilliant red, orange, pink, lavender, or white, rarely with markings of other colors aborally, orally often paler. Highly variable. *H. leviuscula* is not found in Alaska, but there is a complex of several very similar species in Alaska. *H. Leviuscula* is a more southerly species.
SIZE: R to 12 cm.
DISTRIBUTION: Aleutian Islands, Kodiak Island, Prince William Sound, Alaska, to California. DEPTH: 0-435 m.
(Gotshall 1994, Lambert 2000, Clark 2006)

Spiny henricia

Henricia multispina

DESCRIPTION: Medium-sized star; disc small, arms long, tapering to pointed tips. Dorsal plates small, close-set, profusely covered with minute spicules, giving a smoothish appearance to the unaided eye, velvety under magnification. Cannot be confused with any other species. Color uniform whitish or lavender aborally, disc darker with a small red spot at center, white orally.
SIZE: R to 10 cm.
DISTRIBUTION: Bering Sea, Aleutian Islands, British Columbia to Oregon. DEPTH: 100-200 m.
(Clark 2006)

Tumid henricia

Henricia tumida

DESCRIPTION: Small star; arms short and thick; disc swollen. Clumps of spinelets small, close-set; oral rows of pores usually distinguishable. Color highly variable, uniformly colored, pink or light cream or frequently mottled with red, orange, yellow, purple, or brown.
SIZE: R to 3.5 cm.
DISTRIBUTION: Bering Sea, along Alaska Peninsula to Gulf of Alaska. DEPTH: 0-60 m.
(Kessler 1985, Clark 2006)

Henricia cf. *leviuscula*

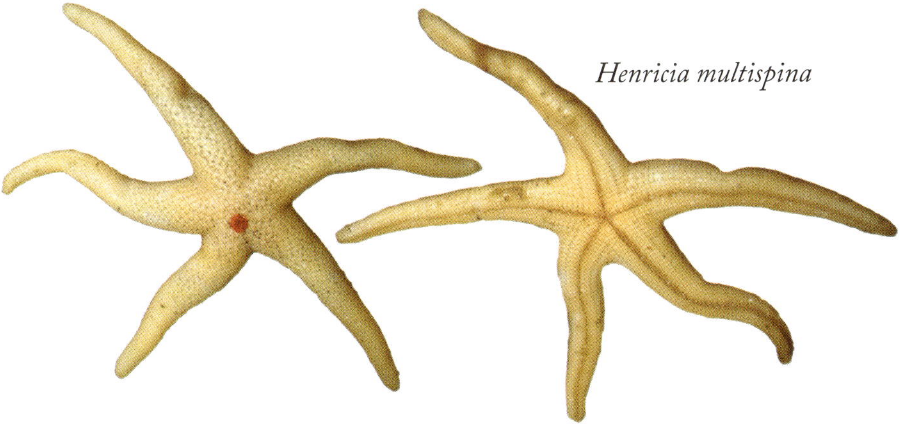

Henricia multispina

Henricia tumida

Bering henricia
Henricia beringiania

DESCRIPTION: Medium-sized star; disc relatively broad, arms moderately long, robust, often inflated. Dorsal plates small, close-set, bearing 2-9 stout spines. Color reddish-brown or orange tones, often mottled aborally, cream orally.
SIZE: R to 5 cm.
DISTRIBUTION: Bering Sea and Aleutian Islands. DEPTH: 40-200 m.
(Clark 2006)

Fuzzy henricia
Henricia sp.

DESCRIPTION: Small star; disc small, arms long. Dorsal plates small, round, relatively close-set with tufts of 8-15 moderately long, thorn-tipped spines. White aborally and orally.
SIZE: R to 2.5 cm.
DISTRIBUTION: Western Aleutian Islands. DEPTH: 200-300 m.
(Clark 2006)

Purple toothed henricia
Odontohenricia sp.

DESCRIPTION: Large star; disc relatively small, arms tapering to slender tips. Dorsal plates small, close-set, bearing 22-36 fine, short spinelets. Color shades of purple or violet aborally and orally; sometimes paler on arms and often with large reddish blotch on disc.
SIZE: R to 15 cm.
DISTRIBUTION: Aleutian Islands. DEPTH: 100-200 m.
(Clark 2006)

Henricia beringiania

Henricia sp.

Odontohenricia sp.

Pale toothed henricia
Odontohenricia sp.

DESCRIPTION: Large star; disc relatively small, arms broad at base, tapering to slender tips. Dorsal plates with 20 to 40 short spinelets. Apical tooth large, robust, pointed. Color pinkish or yellowish-tan with large purple-red blotch on disc aborally, pale tan orally.
SIZE: R to 15 cm.
DISTRIBUTION: Aleutian Islands. DEPTH: 100-200 m.
(Clark 2006)

OTHER HENRICIAS. ITIS (2008) lists 18 species and numerous subspecies of *Henricia*; some forms are intergrades that are hard to identify.

Purple-orange star
Asterias amurensis

DESCRIPTION: Large star; disc broad, 5 arms of moderate length, with broad base, tapering to a point. Flat ventral surface with conspicuous boundary between aboral and oral sides (1). Oral side flat; aboral side roughly textured with profuse, small spines. Color uniformly purple or orange, with white or pale orange spines.
SIZE: R to 20 cm.
DISTRIBUTION: Arctic Ocean to northern Gulf of Alaska. DEPTH: 0-220 m.
(Lambert 2000, Clark 2006)

Giant star
Evasterias echinosoma

DESCRIPTION: Very large star; disc relatively small, arms long, robust. Aboral spines large, uniform, well-spaced, subconical, arranged in a definite longitudinal series along aboral ridge. Color, shape, and spination consistent; upper rows of spines are on side of ray. Color uniform light purple aborally, cream orally.
SIZE: R to 35 cm.
DISTRIBUTION: Chukchi Sea, Bering Sea, and Aleutian Islands.
DEPTH: 25-100 m.
(Kessler 1985, Clark 2006)

Odontohenricia sp.

Asterias amurensis

Evasterias echinosoma

Mottled star

Evasterias troschelii

DESCRIPTION: Very large star; disc small, arms long, usually slender. Spines not uniform in size and shape, not arranged in a definite longitudinal series; color, shape, and spination variable. Color uniform or mottled, gray, purple, orange, pink, tan, bluish, or blue-green aborally and orally.
SIZE: R to 30 cm.
DISTRIBUTION: Bering Sea, Pribilof Islands to California. DEPTH: 0-71 m.
(Kessler 1985, Lambert 2000, Clark 2006)

Arctic star

Leptasterias arctica

DESCRIPTION: Small star; disc relatively broad, arms moderately short, slender to inflated or robust. Spines coarse, knob-like, unequal in diameter. Color variable, shades of brown, orange, and black predominating, spines often cream or tan.
SIZE: R to 6.5 cm.
DISTRIBUTION: Arctic Ocean, Bering Strait, Nunivak Island, to Aleutian and Shumagin islands. DEPTH: 0-80 m.
(Kessler 1985, Clark 2006)

Aleutian star

Leptasterias hylodes

DESCRIPTION: Small star; disc small, arms somewhat short to moderately long, slender to inflated. Aboral spines uniform, delicate, close-set, heavily sheathed in pedicellariae. Color uniform purple brown, tan or white aborally and orally; spines white.
SIZE: R to 5 cm.
DISTRIBUTION: Bering Sea, Aleutian Islands to southeastern Alaska. DEPTH: 1-80 m.
(Kessler 1985, Clark 2006)

OTHER FIVE-RAYED SPECIES OF *LEPTASTERIAS*. The Greenland sea star *L. groenlandica* has spines on the transverse ridges separated by narrow open areas. The sheathed sea star *L. stolocantha* is distinguished by the flat-topped conical sheath of pedicellariae that surrounds the spines, allowing only the tip to protrude.

Evasterias troschelii

Leptasterias arctica

Leptasterias hylodes

Knobby six-rayed star

Leptasterias polaris

DESCRIPTION: Very large star; disc broad; 6 long and robust arms. Prominent cone-like knobs (1); arms and disc variable in shape, knobs variable in prominence. Color variable, uniform or mottled with purple, pink, brown, and orange.
SIZE: R to 25 cm.
DISTRIBUTION: Arctic Ocean, Chukchi Sea, Bering Sea, Aleutian Islands to northwest Gulf of Alaska. DEPTH: 10-100 m.
(Kessler 1985, Clark 2006)

Giant Aleutian six-rayed star

Leptasterias truculenta

DESCRIPTION: Very large 6-rayed star; disc small, arms long and relatively slender, tapering to a point at tips; aboral surface covered with small, widely spaced pointed, conical purple spines (1). Color variable, mottled with red, pink, purple, brown, or white aborally, cream orally.
SIZE: R to 25 cm.
DISTRIBUTION: Aleutian Islands. DEPTH: 25-110 m.
(Clark 2006)

OTHER SIX-RAYED SPECIES OF *LEPTASTERIAS*. The knobless 6-rayed star *L. hexactis* is relatively small and lacks prominent cone-like knobs. In addition, ITIS (2008) lists 20 species and numerous subspecies of *Leptasterias*; some forms are intergrades that are hard to identify.

Sunflower star

Pycnopodia helianthoides

DESCRIPTION: Very large star; disc very broad, arms short, inflated, up to 24 arms or more in large stars. Aboral surface very soft, fleshy, flexible, bearing spines and pedicellariae. Numerous crowded tube feet. Color variable, purple to brown, orange, or yellow.
SIZE: R to 50 cm.
DISTRIBUTION: Eastern Aleutian Islands to California. DEPTH: 0-435 m.
(Gotshall 1994, Lambert 2000, Clark 2006)

Leptasterias polaris

Leptasterias truculenta

Pycnopodia helianthoides

Redbanded star

Orthasterias koehleri

DESCRIPTION: Large star; disc small, arms may be long and slender or coarse and heavy; each aboral spine surrounded by a wreath of pedicellariae. Large, prominent, uniform, light-colored spines (1). Color alternating bands of white and pink, red, or purple, occasionally solid brown.
SIZE: R to 25 cm.
DISTRIBUTION: Eastern Aleutian Islands to California. DEPTH: 1-230 m.
(Kessler 1985, Clark 2006)

Blackspined star

Lethasterias nanimensis

DESCRIPTION: Very large star; disc small; arms 5 or more, long, relatively slender, tapering. Aboral surface evenly covered with short, blunt, black spines (1). Easily broken skeleton. Color tan to dark brown.
SIZE: R to 35 cm.
DISTRIBUTION: Bering Sea, eastern Aleutian Islands to California.
DEPTH: 1-100 m.
(Kessler 1985, Clark 2006)

Long-rayed star

Stylasterias forreri

DESCRIPTION: Very large star; disc small, arms very slender, tapering, fragile at base. Aboral surface with longitudinal rows of long, slender, tapering white spines arising from fleshy mounds (1). Color brown or black with tan mounds.
SIZE: R to 33 cm.
DISTRIBUTION: Eastern Aleutian Islands, southeastern Alaska to California.
DEPTH: 5-530 m.
(Lambert 2000, Clark 2006)

Orthasterias koehleri

Lethasterias nanimensis

Stylasterias forreri

Ochre star

Pisaster ochraceus

DESCRIPTION: Large star, usually 5 thick arms, but the number varies from 4 to 7. Numerous small white spines on the aboral surface arranged in a reticular pattern. Color varies from gray to orange.
SIZE: R to 25 cm.
DISTRIBUTION: Prince William Sound, Alaska, to Baja California.
DEPTH: Above low-tide zone to 90 m.
(Gotshall 1994)

OTHER SEA STARS. The waters of Alaska and especially the Aleutian Island chain harbor the richest region of sea stars in the world. Well over 100 named species in more than 45 genera, including at least 25 recently discovered species, are known to occur in Alaska waters. For more on sea stars, see Clark (2006).

BASKET STAR

Basket star

Gorgonocephalus eucnemis

DESCRIPTION: Very large branching star. Radial shields and arms smooth or with sparse granules; disc large and broad. Arms very long, dichotomously branching into many tendrils. Color pale to dark flesh, pink, or orange-tan on disc, arms paler.
SIZE: Disc diameter to 6 cm; R to 60 cm.
DISTRIBUTION: Bering and Chukchi seas, Gulf of Alaska to California.
DEPTH: 20-2,000 m.
(Kozloff 1987, Clark 2006)

Pisaster ochraceus

Gorgonocephalus eucnemis

SERPENT AND BRITTLE STARS

EXTERNAL FEATURES OF SERPENT AND BRITTLE STARS

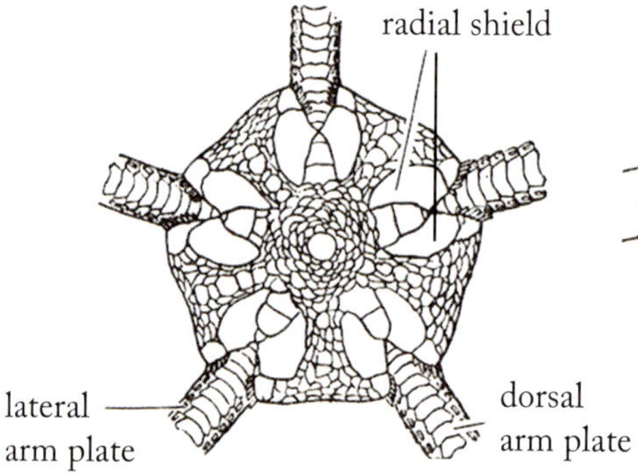

Aboral (dorsal)

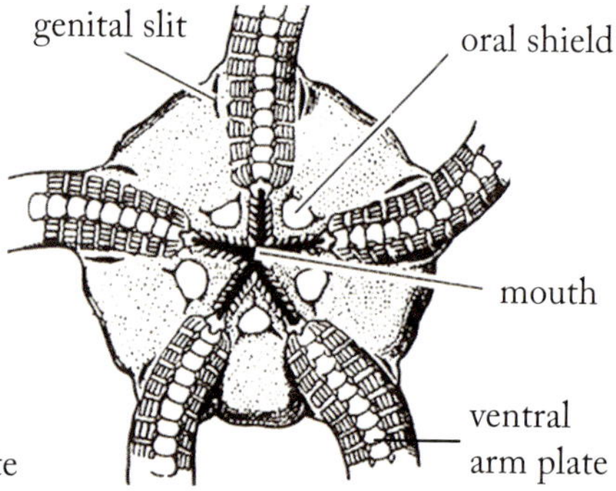

Oral (ventral)

Adapted from Spencer and Wright 1966

Dwarf smooth serpent star

Astronebris tatafilius

DESCRIPTION: Small, smooth star; disc relatively large, inflated or flattened. Arms very long, slender, with prominent groove along the aboral side (1). Color variable, pale orange, red-orange, yellow, pink, or flesh.
SIZE: R to 3.5 cm.
DISTRIBUTION: Aleutian Islands. DEPTH: 80-300 m.
(Clark 2006)

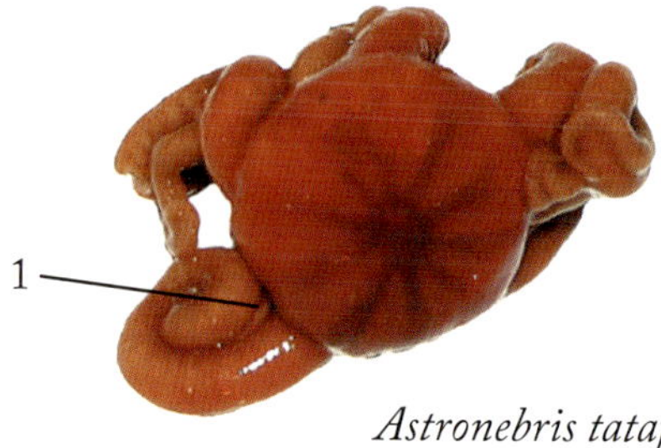

Astronebris tatafilius

Giant serpent star

Asteronyx loveni

DESCRIPTION: Very large star with very long arms capable of coiling around objects for support. Disc large and smooth, inflated; bursal slits very short. Color red-orange to tan, lighter on oral side.

SIZE: R to 35 cm.

DISTRIBUTION: Circumboreal, Aleutian Islands to Bering Sea, southeastern Alaska to California. DEPTH: 100-4,721 m.

(Clark 2006, Lambert and Austin 2007)

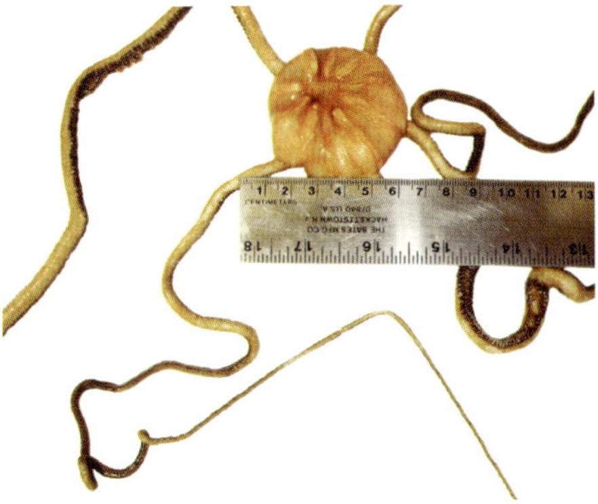

Asteronyx loveni

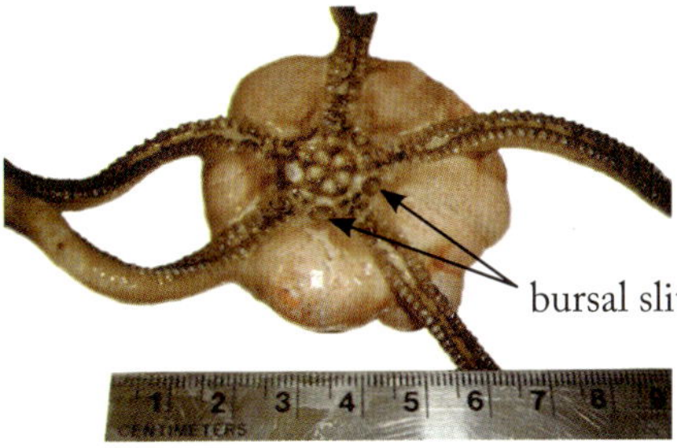

White serpent star

Astrochele laevis

DESCRIPTION: Medium-sized, granular star; disc large, inflated, evenly covered with low granules. Arms long, slender; radial shields prominent. Color white.

SIZE: R to 7 cm.

DISTRIBUTION: Aleutian Islands. DEPTH: 50-200 m.

(Clark 2006)

Astrochele laevis

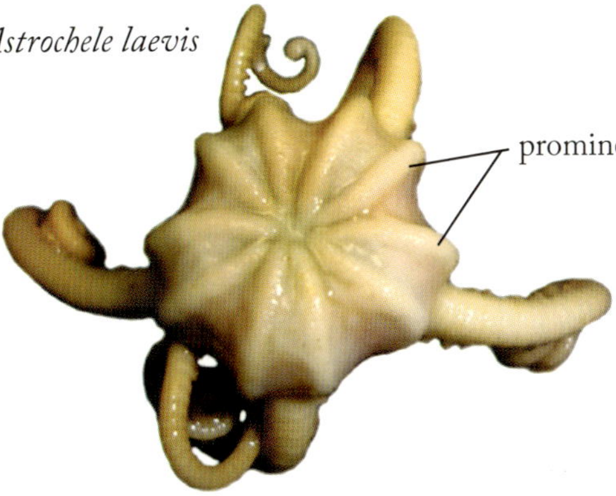

Daisy brittle star

Ophiopholis aculeata kennerlyi

DESCRIPTION: Medium-sized brittle star; disc relatively large, covered with papillae. Radial shields obscured; arms moderately long, flexible. Arm spines short, blunt, about 1.5 times longer than arm plate length; dorsal plates separated by tiny encircling platelets (1). Color highly variable, often mottled or banded with green, red, black, white, and other colors.
SIZE: R to 8 cm.
DISTRIBUTION: Bering Sea, Aleutian Islands, southeastern Alaska to California.
DEPTH: 1-435 m.
(Clark 2006, Lambert and Austin 2007, EOL 2009)

Ubiquitous brittle star

Ophiopholis aculeata aculeata

DESCRIPTION: Medium-sized brittle star; disc fairly large, pentagonal. Small round plates and triangular radial shield obvious (1). Arms long, flexible, dorsal plates separated by tiny encircling platelets; arm spines stout, about twice as long as arm plate length. Color and pattern highly variable, combinations of red, green, brown, white, or orange.
SIZE: R to 8 cm.
DISTRIBUTION: Circumboreal, Bering Sea, southeastern Alaska, and British Columbia. DEPTH: 60-366 m.
(Clark 2006, Lambert and Austin 2007)

Spiny-brown brittle star

Ophiacantha cataleimmoidus

DESCRIPTION: Medium-sized brittle star; disc large, covered with minute granules. Radial shield usually obvious, short, small to elongate, widely separated (1). Arms long; arm spines slender, long, about 4 times as long as arm plate length. Color shades of brown, often with cream to light radial shields and arm bands. (Also known as *Ophiopthalmus cataleimmoidus*.)
SIZE: R to 8 cm.
DISTRIBUTION: Bering Sea, Gulf of Alaska to northern Oregon. DEPTH: 134-1,150 m.
(Clark 2006, Lambert and Austin 2007, EOL 2009)

Ophiopholis aculeata kennerlyi

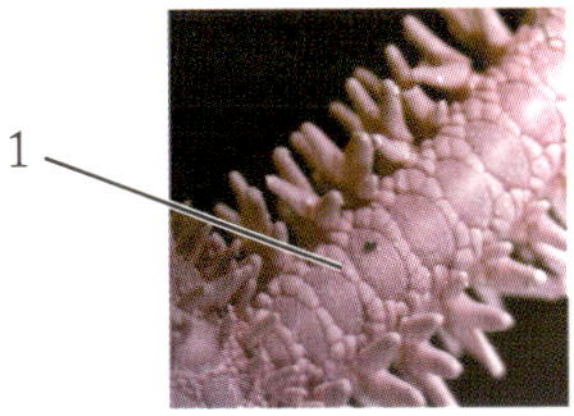

Ophiopholis aculeata aculeata

Ophiacantha cataleimmoidus

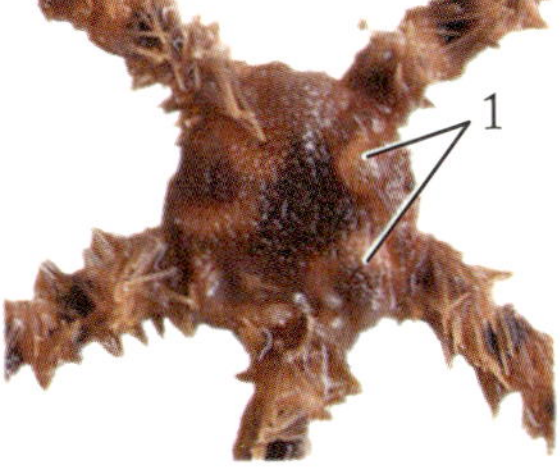

Nine-armed brittle star

Ophiacantha enneactis

DESCRIPTION: Small, 9 to 10 armed brittle star. Disc relatively large, covered with short conical, spine-tipped granules; radial shields and dorsal plates concealed. Arms short, arm spines long. Color pale orange or brown.
SIZE: R to 2.5 cm
DISTRIBUTION: Aleutian Islands. DEPTH: 300-600 m.
(Clark 2006)

Ophiacantha enneactis

Nodulose brittle star

Ophiolebes sp.

DESCRIPTION: Small brittle star; disc large, dorsal plates small and bearing numerous spherical granules. Radial shields small, deeply depressed, triangular (1). Arms short, arm spines short, robust. Color pale orange, lavender, pink, yellow to white. While the three *Ophiolebes* shown here are distinct, species names have not been assigned.
SIZE: R to 3-5 cm.
DISTRIBUTION: Aleutian Islands. DEPTH: 70-500 m.
(Clark 2006)

Ophiolebes sp.

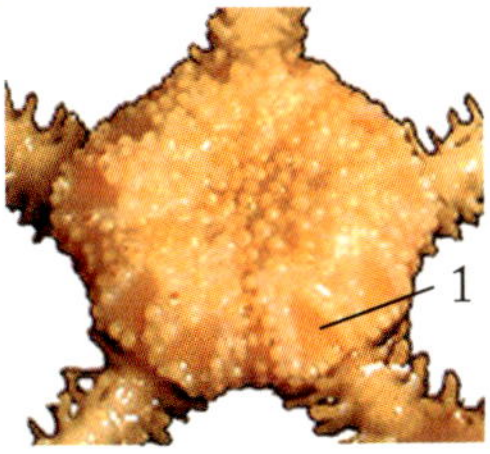

Notched brittle star

Ophiura sarsi

DESCRIPTION: Medium-sized, smooth brittle star. Disc relatively large, with prominent radial shields forming a distinct notch where arms are attached. Largest plates on aboral side of disc separated from each other along entire length by a row of scales. Color green-gray or yellow-gray tones. Both the disc and arms are white orally.
SIZE: R to 8 cm.
DISTRIBUTION: Circumboreal, Arctic, Chukchi, and Bering seas, Aleutians and Kodiak Island, southeastern Alaska south to California. DEPTH: 0-1,460 m.
(Kessler 1985, Clark 2006, Lambert and Austin 2007, EOL 2009)

Great armored brittle star

Stegophiura ponderosa

DESCRIPTION: Large armored brittle star. Disc pentagonal, inflated; dorsal plates like paving stones. Arms rather rigid, triangular in cross section, thick at base, tapering to slender tips. Color reddish or orange. (Also known as *Amphiophiura ponderosa*.)
SIZE: R to 20 cm.
DISTRIBUTION: Aleutian Islands, Kodiak Island, British Columbia to California. DEPTH: 73-1,436 m.
(Clark 2006, Lambert and Austin 2007)

Ophiura sarsi

Stegophiura ponderosa

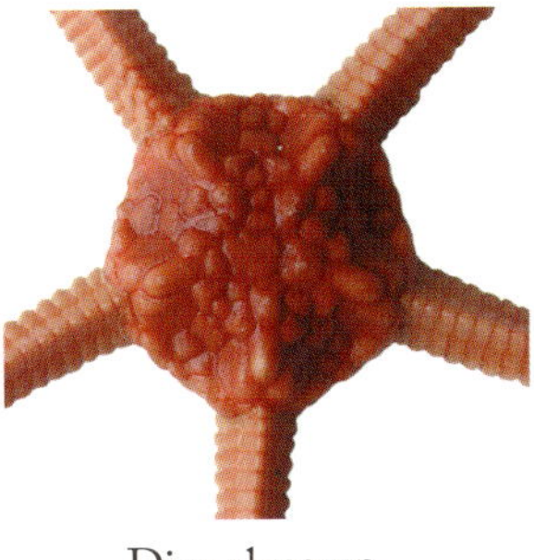

Disc close up

SEA URCHINS

Aleutian urchin

Strongylocentrotus polyacanthus

DESCRIPTION: Medium-sized hemispherical urchin. Spines short, robust. Tissue and tube feet pale green to white; spines usually purple-brown at base (1) becoming green or white at tips; sometimes uniformly pale green.
SIZE: Diameter to 6.5 cm.
DISTRIBUTION: Aleutian Islands to western Gulf of Alaska.
DEPTH: 1-60+ m.
(Clark 2006)

Green urchin

Strongylocentrotus droebachiensis

DESCRIPTION: Medium-sized hemispherical urchin; covered with short stiff spines (1 to 1.5 cm in length), usually greenish or whitish, sometimes dark purple. Secondary spines slightly shorter and finer than primary spines. Tube feet dark greenish or purple.
SIZE: Diameter to 12 cm.
DISTRIBUTION: Circumboreal, Arctic Ocean, Kodiak Island, Cook Inlet, Prince William Sound, British Columbia, Washington. DEPTH: 0-300 m.
(Kessler 1985, Clark 2006, Lambert and Austin 2007, EOL 2009)

Red urchin

Strongylocentrotus franciscanus

DESCRIPTION: Large hemispherical urchin. Spines sharply pointed, very long (about ⅔ as long as the diameter of the test). Tissue dark reddish or purple; spines purple or red.
SIZE: Diameter to 30 cm.
DISTRIBUTION: Kodiak Island, Gulf of Alaska, southeastern Alaska, British Columbia to Baja California. DEPTH: 1-125 m.
(Kessler 1985, Clark 2006, Lambert and Austin 2007)

Strongylocentrotus polyacanthus

Strongylocentrotus droebachiensis

Strongylocentrotus franciscanus

Fragile (pink) urchin

Strongylocentrotus fragilis

DESCRIPTION: Medium-sized hemispherical urchin; test (shell) thin and easily broken. Disc not as thick from top to bottom as *Strongylocentrotus droebachiensis.* Spines short, slender; tips often white; color pinkish-red or pink-orange. (Also known as *Allocentrotus fragilis.*)
SIZE: Diameter to 12 cm.
DISTRIBUTION: Bering Sea, British Columbia to Baja California. DEPTH: 50-1,260 m.
(Kessler 1985, Clark 2006, Lambert and Austin 2007, EOL 2009)

Heart urchin

Brisaster latifrons

DESCRIPTION: Medium-sized, heart-shaped urchin; test roughly heart-shaped, fragile. Aboral side of test with deep groove along the anterior half (1), leading to the mouth. Posterior end rather flattened; posterior ambulacral petals about half the length of anterior petals. Color dull greenish to light brown.
SIZE: Length to 7 cm.
DISTRIBUTION: Aleutian Islands, Gulf of Alaska, southeastern Alaska, British Columbia to California. DEPTH: 51-1,800 m.
(Clark 2006, Lambert and Austin 2007, EOL 2009)

OTHER SEA URCHINS. The white sea urchin *Strongylocentrotus pallidus* can be confused with *S. droebachiensis* but the white color and fewer spines of *S. pallidus* separates them. The purple urchin *S. purpuratus* has stout spines and a bluish-purple color that is fairly distinctive. None of the other sea urchins have a similar color.

SAND DOLLAR

Northern (Parma) sand dollar

Echinarachnius parma

DESCRIPTION: Body flattened, disc-shaped. Disc more or less symmetrical, petals of ambulacral star centered and of equal length (1). Color black to green.
SIZE: Diameter to 7 cm.
DISTRIBUTION: Circumboreal, Bering Sea, Aleutian Islands to Gulf of Alaska. DEPTH: 1-150 m.
(Kessler 1985, Clark 2006, Lambert and Austin 2007, EOL 2009)

Strongylocentrotus fragilis

Brisaster latifrons

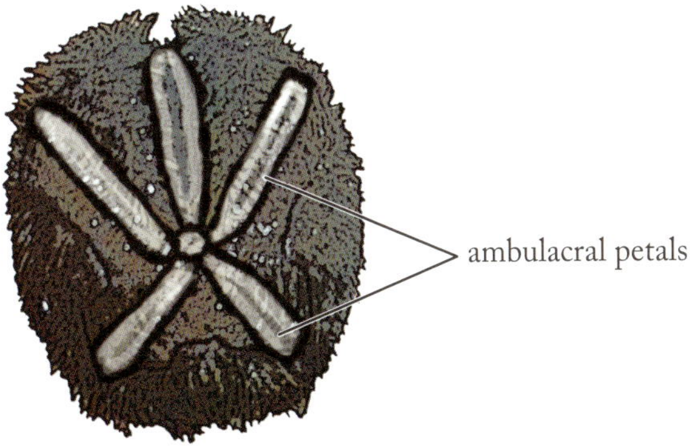

Echinarachnius parma

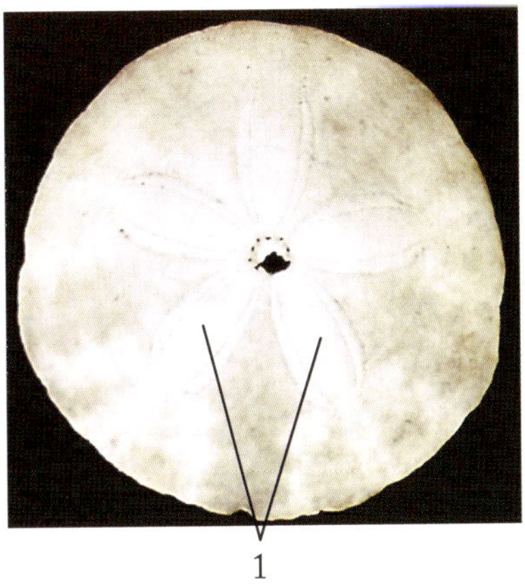

Dried specimen without spines

SEA CUCUMBERS

Giant California cucumber

Parastichopus californicus

DESCRIPTION: Very large, firm, elongate cucumber. Body covered with prominent stiff, conical papillae (1); tube feet on ventral side (2). Color reddish, purple-brown, or yellow, with light-colored papillae.
SIZE: Length to about 50 cm.
DISTRIBUTION: Gulf of Alaska, British Columbia to Mexico. DEPTH: 0-250 m.
(Gotshall 1994, Lambert 1997, Clark 2006)

Northern papillate (slender) cucumber

Bathyplotes sp.

DESCRIPTION: Large elongate, papillose cucumber. This cucumber is very similar to *Synallectes challengeri* (not shown), but is distinguished by its larger size, more slender outline, and papillae that are more slender and widely spaced and appear to be in rows. Color brown to tan tones. Taxonomy of this group is not well documented; field identification should be made to genus only.
SIZE: Contracted length to 30 cm.
DISTRIBUTION: Bering Sea to western Gulf of Alaska. DEPTH: 30-200 m.
(Kessler 1985, Lambert 1997, Clark 2006)

Red-orange armored cucumber

Psolus japonicus

DESCRIPTION: Medium-sized, dome-shaped cucumber; ventral surface flat. Dorsal surface covered with numerous, rather small convex scales. Color red-orange.
SIZE: Length to 10 cm.
DISTRIBUTION: Aleutian Islands to Gulf of Alaska. DEPTH: 30-150 m.
(Clark 2006)

Parastichopus californicus

Bathyplotes sp.

Contracted view

Psolus japonicus

White-scaled cucumber

Psolus squamatus

DESCRIPTION: Medium-sized, low profile, dome-shaped cucumber. Between the tentacles and the anus there are about 12 plates; thin dorsal plates overlap smoothly but have granules along edges. Tube feet form a conspicuous double row around the perimeter of the sole (1) with a few scattered down the midline. Dorsal scales white and smooth, but often discolored brown.
SIZE: Length to 13 cm, normally ranges between 5 and 10 cm.
DISTRIBUTION: Aleutian Islands, Gulf of Alaska, British Columbia to California.
DEPTH: 200-3,500 m.
(Lambert 1997, Clark 2006)

Arctic armored cucumber

Psolus fabricii

DESCRIPTION: Medium-sized, dome-shaped cucumber. Ventral surface flat, dorsal scales not overlapping, surface granular (1). Color red.
SIZE: Length to 7 cm.
DISTRIBUTION: Circumboreal, south to Aleutian Islands. DEPTH: 10-80 m.
(Clark 2006)

Sea football

Cucumaria fallax

DESCRIPTION: Large cucumber; football-shaped when contracted. Tube feet retractable, in longitudinal rows. Color yellowish to pale orange or white, tentacles orange.
SIZE: Contracted length to 30 cm.
DISTRIBUTION: Bering Sea, Aleutian Islands to Gulf of Alaska. DEPTH: 40-150 m.
(Kessler 1985, Lambert 1997, Clark 2006)

Psolus squamatus

Psolus fabricii

Cucumaria fallax

Expanded

Contracted

Crescent cucumber

Pentamera lissoplaca

DESCRIPTION: Very small cucumber, usually less than 2 inches long. Body crescent-shaped; tube feet small and arranged in 5 distinct, crowded double rows. Color white or cream.
SIZE: Length to 5 cm.
DISTRIBUTION: Gulf of Alaska, British Columbia to California.
DEPTH: 50-300 m.
(Kessler 1985, Lambert 1997, EOL 2009)

Sweet sea potato

Molpadia intermedia

DESCRIPTION: Medium-sized cucumber with a tapering "tail" (1). Body elongate, firm, and smooth. Color dark purple, maroon, or reddish-brown.
SIZE: Length to 14 cm.
DISTRIBUTION: Kodiak Island, Gulf of Alaska, southeastern Alaska, British Columbia to Baja California. DEPTH: 7-2,925 m.
(Kessler 1985, Lambert 1997, Clark 2006, EOL 2009)

TUNICATES

Sea onion

Boltenia ovifera

DESCRIPTION: Medium-sized, onion-like stalked tunicate. Body globe-like with 2 siphon-like openings on stalk. Grows singly or in large clumps. Color pale white-tan, pale orange or red-orange.
SIZE: Height including stalk to 30 cm.
DISTRIBUTION: Arctic Ocean, Bering Sea, and Aleutian Islands.
DEPTH: 30-100 m.
(Kessler 1985)

Pentamera lissoplaca

Molpadia intermedia

Boltenia ovifera

Bristly tunicate

Halocynthia hilgendorfi igaboja

DESCRIPTION: Large, globular or oval tunicate, covered with stiff bristles. Color olive-brown to tan.
SIZE: Height to 10 cm.
DISTRIBUTION: Aleutian Islands to California. DEPTH: 1-170 m.
(Clark 2006)

Sea peach

Halocynthia aurantium

DESCRIPTION: Large, globular or sack-like tunicate. Body greatly inflated, with smooth or wrinkled skin; skin texture fuzzy, covered by minute tubercles. Color orange or orange-red.
SIZE: Height to 15 cm.
DISTRIBUTION: Bering Sea, Aleutian Islands, southeastern Alaska to California. DEPTH: 2-100 m.
(Kessler 1985, Clark 2006, EOL 2009)

Sea potato

Styela rustica

DESCRIPTION: Large, potato-shaped tunicate; siphons when contracted are pointy-like tips located on upper end. Usually grows in clumps. Color brown.
SIZE: Height to 10 cm.
DISTRIBUTION: Arctic Ocean, Gulf of Alaska. DEPTH: 10-100 m.
(Kessler 1985, Clark 2006, EOL 2009)

Halocynthia hilgendorfi igaboja

Halocynthia aurantium

Styela rustica

LITERATURE

Arnsberg, A.J. 2001. Arthropoda, Cirripedia: The barnacles. In: A.L. Shanks (ed.), An identification guide to the larval marine invertebrates of the Pacific Northwest. Oregon State University Press, Corvallis.

Barr, L., and N. Barr. 1983. Under Alaskan seas: The shallow water marine invertebrates. Alaska Northwest Publishing Company, Anchorage, Alaska. 280 pp.

Behrens, D.W. 1991. Pacific coast nudibranchs: A guide to the opisthobranchs, Alaska to Baja California. Sea Challengers, Monterey, California. 107 pp.

Bernard, F.R. 1972. The genus *Thyasira* in western Canada (Bivalvia: Lucinacea). Malacologica 11(2):365-389.

Bernard, F.R. 1983. Catalogue of living *Bivalvia* of the eastern Pacific Ocean: Bering Strait to Cape Horn. Can. Spec. Publ. Fish. Aquat. Sci. 61. 102 pp.

Butler, T.H. 1980. Shrimps of the Pacific coast of Canada. Can. Bull. Fish. Aquat. Sci. 202. 208 pp.

California Academy of Sciences. 2009. Digital resource at http://research.calacademy.org/research/ichthyology/catalog/fishcatmain.asp.

Coan, E.V., V. Scott, and F.R. Bernard. 2000. Bivalve sea shells of western North America: Marine bivalve mollusks from arctic Alaska to Baja California. Santa Barbara Museum of Natural History, Santa Barbara, California. 764 pp.

Clark, R. 2006. Field guide to the benthic marine invertebrates of Alaska's shelf and upper slope. National Oceanic and Atmospheric Administration, NMFS AFSC unpublished manuscript. 305 pp.

Donaldson, W.E., and S.C. Byersdorfer. 2005. Biological field techniques for lithodid crabs. Alaska Sea Grant College Program, University of Alaska Fairbanks. 76 pp.

Encyclopedia of Life (EOL). 2009. Digital resource at http://www.eol.org.

Foster, N.R. 1991. Intertidal bivalves: A guide to the common marine bivalves of Alaska. University of Alaska Press, Fairbanks, Alaska. 152 pp.

Gotshall, D.W. 1994. Guide to marine invertebrates, Alaska to Baja California. Sea Challengers, Monterey, California. 105 pp.

Gotshall, D.W., and L.L. Laurent. 1979. Pacific Coast subtidal marine invertebrates: A fishwatcher's guide. Sea Challengers, Monterey, California. 107 pp.

Hall, S., and S. Thatje. 2009. Four new species of the family Lithodidae (Decapoda: Anomura) from the collections of the National Museum of Natural History, Smithsonian Institution. Zootaxa 2302:31-47.

Harbo, R.M. 1997. Shells and shellfish of the Pacific Northwest: A field guide. Harbour Publishing Co., Madeira Park, B.C., Canada. 270 pp.

Hart, J.F. 1982. Crabs and their relatives of British Columbia. British Columbia Provincial Museum, Victoria, B.C., Canada. 267 pp.

Integrated Taxonomic Information System (ITIS). 2008. Digital resource at http://www.itis.gov.

Jadamec, L.S., W.E. Donaldson, and P. Cullenberg. 1999. Biological field techniques for *Chionoecetes* crabs. Alaska Sea Grant College Program, University of Alaska Fairbanks. 80 pp.

Jensen, G.C. 1995. Pacific coast crabs and shrimps. Sea Challengers, Monterey, California. 87 pp.

Johnson, M.E., and H.J. Snook. 1955. Seashore animals of the Pacific Coast. Dover Publications, Inc., New York. 659 pp.

Jorgensen, E.M. 2009. Field guide to squids and octopods of the eastern North Pacific and Bering Sea. Alaska Sea Grant College Program, University of Alaska Fairbanks.

Keen, A.M., and E. Coan. 1974. Marine molluscan genera of western North America: An illustrated key. Stanford University Press, Stanford, California. 208 pp.

Kessler, D.W. 1985. Alaska's saltwater fishes and other sea life: A field guide. Alaska Northwest Publishing Company, Anchorage, Alaska. 358 pp.

Kozloff, E.N. 1983. Seashore life of the northern Pacific Coast: An illustrated guide to northern California, Oregon, Washington, and British Columbia. University of Washington Press, Seattle, Washington. 370 pp.

Kozloff, E.N. 1987. Marine invertebrates of the Pacific Northwest. University of Washington Press, Seattle, Washington. 511 pp.

Kramer, D.E., and V.M. O'Connell. 2003. Guide to northeast Pacific rockfishes: Genera *Sebastes* and *Sebastolobus*. Alaska Sea Grant College Program, University of Alaska Fairbanks. 78 pp.

Kramer, D.E., W.H. Barss, B.C. Paust, and B.E. Bracken. 1995. Guide to northeast Pacific flatfishes: Families Bothidae, Cynoglossidae, and Pleuronectidae. Alaska Sea Grant College Program, University of Alaska Fairbanks. 104 pp.

Lamb, A., and P. Edgell. 1986. Coastal fishes of the Pacific Northwest. Harbour Publishing Co., Madeira Park, B.C., Canada. 224 pp.

Lambert, P. 1997. Sea cucumbers of British Columbia, southeast Alaska and Puget Sound. Royal British Columbia Museum Handbook. University of British Columbia Press, Victoria, B.C., Canada. 166 pp.

Lambert, P. 2000. Sea stars of British Columbia, southeast Alaska and Puget Sound. Royal British Columbia Museum Handbook, University of British Columbia Press, Toronto, Canada. 186 pp.

Lambert, P., and W. Austin. 2007. Brittle stars, sea urchins and feather stars of British Columbia, southeast Alaska and Puget Sound. Royal British Columbia Museum Handbook. University of British Columbia Press, Victoria, B.C., Canada. 150 pp.

MacIntosh, R.A. 1976. A guide to some common eastern Bering Sea snails. NOAA NMFS Processed Report, Kodiak, Alaska. 27 pp.

Makarov, V.V. 1962. Fauna of U.S.S.R., Crustacea, Vol. X, No. 3. S.A. Zernov and A.A. Shtakel'berg (eds.). Israel Program for Scientific Translations, Jerusalem. 283 pp.

Marine Species. 2009. World Asteroidea Database digital resource at http://www.marinespecies.org/asteroidea.

McLaughlin, P.A. 1972. The hermit crabs of the genus *Pagurus* (Crustacea: Decapoda: Paguridae) from northwestern North America, with a partial revision of the genus. Ph.D. thesis, George Washington University, Washington, D.C. 554 pp.

McLaughlin, P.A. 1974. The hermit crabs (Crustacea: Decapoda: Paguridae) of northwestern North America. Zool. Verhandel. Leiden 130. 396 pp.

McLaughlin, P.A., et al. 2005. Common and scientific names of aquatic invertebrates from the United States and Canada: Crustaceans. American Fisheries Society, Special Publication 31, Bethesda, Maryland.

Mecklenburg, C., T.A. Mecklenburg, and L.K. Thorsteinson. 2002. Fishes of Alaska. American Fisheries Society, Bethesda, Maryland. 1037 pp.

Miller, D.J., and R.N. Lea. 1972. Guide to the coastal marine fishes of California. California Department of Fish and Game, Sacramento, Fishery Bulletin 157. 235 pp.

Morris, P.A. 1966. A field guide to Pacific Coast shells, including shells of Hawaii and the Gulf of California, 2nd edn. Peterson Field Guide Series. Houghton Mifflin Company, Boston. 297 pp.

O'Clair, R.M., and C.E. O'Clair. 1998. Southeast Alaska's rocky shores: Animals. Plant Press, Auke Bay, Alaska. 564 pp.

Orr, J.W., and J.E. Blackburn. 2004. The dusky rockfishes (Teleostei: Scorpaeniformes) of the North Pacific Ocean: Resurrection of *Sebastes variabilis* (Pallas, 1814) and a redescription of *Sebastes ciliatus* (Tilesius, 1813). Fish. Bull. U.S. 102:328-348.

Orr, J.W., and M.S. Busby. 2006. Revision of the snailfish genus *Allocareproctus* Pitruk & Fedorov (Teleostei: Liparidae), with descriptions of four new species from the Aleutian Islands. Zootaxa 1173:1-37.

Orr, J.W., and S. Hawkins. 2008. Species of the rougheye rockfish complex: Resurrection of *Sebastes melanostictus* (Matsubara, 1934) and a redescription of *Sebastes aleutianus* (Jordan and Evermann, 1898) (Teleostei: Scorpaeniformes). Fish. Bull. U.S. 106:111-134.

Orr, J.W., and K.P. Maslenikov. 2007. Two new variegated snailfishes of the genus *Careproctus* (Teleostei: Scorpaeniformes: Liparidae) from the Aleutian Islands, Alaska. Copeia 2007(3):699-710.

Rice, T. 1973. Marine shells of the Pacific Coast. Erco, Inc., Tacoma, Washington. 102 pp.

Smirnov, A.V. 1982. Species *Pteraster obscurus* (Perrier, 1891) and *P. octaster* Verrill, 1909 (Asteroidea: Pterastericae). Explorations of the Fauna of the Seas 29(37):102-113. (In Russian with English summary.)

Spencer, W.K., and C.W. Wright. 1966. Asterozoans. Treatise on invertebrate paleontology, part U, Echinodermata 3, vol. 1, no. U4-U107. Geological Society of America and University of Kansas Press.

Squires, H.J. 1992. Recognition of *Pandalus eous* Makarov, 1935, a Pacific species, not a variety of the Atlantic *Pandalus borealis* Kroyer, 1838 (Decapoda, Caridea). Crustaceana 63(3):257-262.

Stevens, B.G. 2002. Checklist of Alaskan crabs. In: A.J. Paul, et al. Crabs in cold water regions: Biology, management, and economics. Alaska Sea Grant College Program, University of Alaska Fairbanks, pp. 5-8.

Stevenson, D.E., J.W. Orr, G.R. Hoff, and J.D. McEachran. 2007. A field guide to sharks, skates, and ratfish of Alaska. Alaska Sea Grant College Program, University of Alaska Fairbanks. 77 pp.

Turgeon, D.D., et al. 1998. Common and scientific names of aquatic invertebrates from the United States and Canada: Mollusks, 2nd edn. American Fisheries Society, Special Publication 26, Bethesda, Maryland.

Wing, B.W., and D.R. Barnard. 2004. A field guide to Alaskan corals. NOAA Tech Memo, NMFS-AFSC-146. 67 pp.

Zaklan, S.D. 2002. Review of the Lithodidae (Crustacea: Anomura: Paguroidea): Distribution, biology, and fisheries. In: A.J. Paul, et al. (eds.), Crabs in cold water regions: Biology, management, and economics. Alaska Sea Grant College Program, University of Alaska Fairbanks, pp. 751-845.

PHOTO TECHNIQUES AND CREDITS

Most of the photographs in this book were taken over a period of ten years while biologists were on surveys in the Bering Sea, Aleutian Islands, St. Matthew and Pribilof island areas, and Gulf of Alaska, using an Olympus®E10 digital camera and an Olympus®Evolt E-330 SLR digital camera. All photos were processed using Adobe Photoshop®CS2 to improve clarity by removing the background. Product names used in this book are included for scientific completeness, but do not constitute a product endorsement. The majority of photographs were taken by Susie Byersdorfer and other ADFG personnel on yearly pot and trawl surveys.

PHOTO CREDITS

Aaron Baldwin, University of Alaska Fairbanks, Juneau, Alaska
fragile whelk

Ryan Burt, Alaska Department of Fish and Game (ADFG), Kodiak, Alaska
"leopard" and mud skates, rock greenling, armorhead and Leister sculpins, blotched snailfish, soccer ball sponge, fuzzy cratered sponge, furry ball sponge, baseball sponge, Bering chiton, "spotted" diaulula, Dall's dentronotid, gravel isopod, spiky and spiny arcturids, skeleton shrimp, scaled crab, hairy vase and cotton ball sponges, spiny lithode and longhorn decorator crabs, arctic sand star, red cookie star, beautiful sun star, tesselated slime star, crested star, spiny henricia, Bering henricia, fuzzy henricia, giant Aleutian six-rayed star, white serpent star, daisy brittle star, nodulose brittle star, nine-armed brittle star, northern papillate (slender) cucumber.

Roger Clark, Insignis Biological Consulting, Eagle Mountain, Utah
obscure cushion star, tumid henricia, stellate cookie star, red cookie star, eight-rayed cushion star, discordant mussel.

Matt Foster, ADFG, Kodiak, Alaska
cutthroat trout, chum salmon.

Dave Jackson, ADFG, Kodiak, Alaska
sturgeon poacher (close-up), decorated warbonnet, humpy shrimp.

Wes Jones, Norton Sound Economic Development Corporation, Nome, Alaska
arctic lamprey.

Rachel Latham, ADFG, Kodiak, Alaska
chinook salmon, quillback and harlequin rockfishes, dock shrimp, Alaska eelpout, Pilsbry piddock.

Heidi Morrison, ADFG, Kodiak, Alaska
butterfly crab.

National Marine Fisheries Service, Seattle, Washington
triangle Tanner crab.

Larry Neff, ADFG, Nome, Alaska
Bering wolffish (juvenile).

Verlin Pherson, Scuba Do, Kodiak, Alaska
giant orange tochui, golden dirona.

Laura Slater, ADFG, Kodiak, Alaska
snow crab, Alaska pink shrimp.

Kally Spalinger, ADFG, Kodiak, Alaska
disjunct cleft clam, Greenland smoothcockle, whitebarred prickleback, saffron cod.

Brad Stevens, National Oceanic and Atmospheric Administration (NOAA), Kodiak, Alaska
Verill's paralomis.

Patrick Sullivan, University of Alaska Anchorage, Anchorage, Alaska
armorhead sculpin (side).

Gary Todd, ADFG, Nome, Alaska
polar eelpout, blackline prickleback (close-up).

Donn Tracy, ADFG, Kodiak, Alaska
rainbow trout, steelhead, Dolly Varden, coho salmon (spawning).

Dan Urban, NOAA, Kodiak, Alaska
Kamchatka flounder.

Mike Vaughn, ADFG, Sitka, Alaska
yelloweye rockfish, juvenile.

Leslie Watson, ADFG, Kodiak, Alaska
rainbow smelt, brightbelly sculpin, polar eelpout (adult), blackline prickleback (whole), soft crab, giant star.

Carrie Worton, ADFG, Kodiak, Alaska
China rockfish, tiger rockfish, yelloweye rockfish, rhinoceros crab.

GLOSSARY

Aboral: In sea stars and brittle stars, the dorsal side of the animal with the anus.

Adipose fin: A fleshy fin, without rays and usually small, on the dorsal median line of the body, in front of the tail.

Aeolid: Type of nudibranch possessing cerata rather than a plume-like gill on the back and no rhinophoral sheath.

Ambulacral groove: In sea stars, a ventral groove or furrow that runs lengthwise down the arm, roofed by a series of ambulacral ossicles.

Ambulacral: In echinoderms, applies to those areas of the body bearing tube feet.

Anadromous: Ascending rivers from the sea for breeding, as some species of marine fishes.

Anal: Near or pertaining to the anus.

Anal fin: An unpaired fin on the ventral side of the body between the anus and the tail, in fish.

Anal fin spine: A hard, sharp spine pointing anterior and located anterior to the anal fin of a fish.

Antenna: Anterior jointed sensory appendage.

Anterior: On or toward the front or head end.

Anus: Opening at the lower end of the alimentary canal through which solid waste is expelled.

Aperture: In snails and gastropods, the opening from which the soft body extends from the shell.

Apical tooth: In sea stars and brittle stars, a small plate or spine attached to a plate on the inner edge of the jaw.

Arm: In feather stars, sea stars, and brittle stars, a moveable projection distal to the disc, sometimes called a ray.

Axial: Approximately parallel to the axis of coiling in snail shells.

Axial rib (or fold): In gastropod shells, a ridge running longitudinally on the shell (across the whorls); nearly perpendicular to the whorls and to spiral ridges (see p. 146).

Axial sculpture: Shell relief pattern that follows the axis of the shell (see p. 146).

Barbel: Elongate, fleshy appendage or projection on the snout, around the mouth, or below the lower jaw, in fish.

Basal: Relating to or arising from a base.

Beak: Located above the hinge, the first-formed part of a bivalve shell (see p. 185). Also called the umbo. In barnacles, the umbo is found at the base of the shell.

Benthic: At the bottom of a body of water.

Blind side: The bottom side of a flatfish; lacks eyes and usually has little or no pigmentation.

Branchial plume: The respiratory structure (gill) of a nudibranch usually on the posterior end of the dorsum.

Branchiostegal rays: Bony or cartilaginous supports of the gill membranes, located on the underside of the head, below the opercular bones and behind the lower jaw, and attached to the hyoid arch, in fish.

Bursal: Referring to bursa, one of a pair of pouches in the disc of brittle stars, adjacent to each arm. Gonads empty eggs or sperm into the bursa before release via a hole or slit to the outside.

Bushy: Coral colonies with abundant branches arising immediately above the holdfast and not forming an obvious stem.

Calcareous: Of, containing, or like calcium carbonate.

Carapace: Exoskeleton covering head and thorax (cephalothorax), in crustaceans.

Carapace length: The straight-line distance across the carapace from the posterior margin of the right eye orbit to the medial-posterior margin of the carapace.

Carapace width: The greatest straight-line distance across the carapace excluding spines.

Carpus: Fifth article or segment of an appendage, of a crab.

Caudal: At or near the tail or hind parts; posterior. In this book it describes fish.

Caudal peduncle: The narrow fleshy end of the body that supports the caudal fin.

Ceras (pl. cerata): Finger-like respiratory structure bearing extensions of the digestive gland and occurring in groups of parallel series along the dorsum of some nudibranchs.

Chalice: The cup-shaped skeletons of some scleractinian corals.

Chela (pl. chelae): A pincer-like claw of a crustacean, such as a crab.

Cheliped: Modified crab leg with a pincer or claw at the distal end.

Chondrophore: In bivalves, shelf-like, spoon-like, or socket-like part of the hinge.

Circumboreal: Found in the cool water regions south of the Arctic; including the Atlantic and Pacific Oceans.

Cirrus (pl. cirri): A slender tentacle-like appendage or fringe. In this book it describes fish.

Clasper: A reproductive structure located along the posterior margin of the pelvic fins in male sharks and skates.

Claw: Sharp corneous tip of dactyl of walking legs, in crabs.

Cloaca: The combined opening of the digestive and urogenital system.

Columella: The axis or central pillar of a spiral shell.

Commensal: The relationship between one animal (or plant) living harmlessly with another and sharing its food.

Contractile polyp: A polyp that can diminish in size without introversion or retraction, in corals.

Coronal: Pertaining to the upper portion of the head or crown.

Coronal spine: Spine between the frontal and parietal bones of the head.

Crenate (or crenulate): A margin of small, rounded, scallop-like projections.

Cyclosystems: Cluster of hydrocoral polyps in which the gastropore is surrounded by a group of dactylopores.

Dactyl: Finger or toe, applied to the distal joint of the crustacean leg.

Dactylopore: Small pore housing defensive hydrocoral polyps.

Demersal: Referring to organisms that live near the bottom of the ocean.

Denticle: The hooked, tooth-like scales found on sharks and skates.

Dextral: In snails, the shell coils or turns from left to right, right-handed; opposite of sinistral. Most gastropods have shells with a right-handed (dextral) coil; few have a left-handed (sinistral) coil.

Dichotomous: Divided into two parts.

Disc: The round or pentagonal central body region of sea stars, basket stars, and brittle stars.

Distal: Farthest from the body (opposite of proximal).

Dorsal: On or toward the back (dorsum); opposite of ventral.

Dorsal fin: An unpaired fin on the dorsal or upper side of the body between the head and the tail, on a fish.

Dorsal polygonal plate: The dorsal (aboral) surface of each species of sea star has a characteristic pattern of plates, spines, and granules. Polygonal plates have more than four sides.

Dorsal spine: Hard, usually sharp fin supports found in the anterior part of the dorsal fin in some fishes.

Endopodite: In crustaceans the inner branch of a biramous appendage.

Epibenthic: Located on the bottom of a water body, as opposed to in the bottom.

Epipelagic: Living in the upper part of the ocean, from the surface to about 200 m, or in the sunlit depths.

Eversible: Capable of being turned outward.

Evert: To turn outward or inside out.

Exopodite: In crustaceans the outer branch of a biramous appendage.

Eyed side: The top side of a flatfish. This side has eyes and is usually more darkly pigmented than the side that lies against the bottom.

Flammule: Flame-shaped color pattern.

Fusiform: Spindle-shaped; tapering toward both ends, but more abruptly anteriorly, with sides flattened only slightly or not at all. In this book, it describes whelks.

Galatheid crabs: Crabs that belong to the family Galatheidae. They are characterized by having four pairs of visible legs (first legs long) and a shrimp-like body.

Gastropore: Pores in the hard skeleton of a hydrocoral that accommodate feeding polyps.

Gill raker: A bony, tooth-like structure on the anterior edge of the gill arch used for protection or for straining out food, in fish.

Girdle: In chitons, the muscular tissue that holds the plates together and forms the margin.

Gorgonian corals: Corals having a horny, calcareous branching skeleton.

Hinge ligament: In bivalves, the ligament that springs the shells open when the adductor muscles relax. The external hinge ligaments are found posterior to the umbo (beak) at the siphon end, along the dorsal margin.

Hydrocoral: A colonial hydroid that has a calcareous skeleton, either encrusting on rock and shell or forming an upright colony that is variously branched.

Hydrotheca: In hydroids, the hard covering that surrounds an individual feeding polyp.

Interorbital space: The space on top of the head between the bony edges of the orbits or eye sockets, in fish.

Isthmus: Narrow triangular portion of the body beneath the head of some fish where the gill membranes converge.

Keel: Elevated ridge.

King crab: The largest of all the commercially important crab species. The family Lithodidae to which king crabs belong has 16 genera and 95 known species. King crabs have three pairs of walking legs and a body covered with sharp thornlike spines.

Kype: The hooked jaw developed at spawning time by the males of some salmonid fishes.

Lamellar rib: Prominent protrusion on the surface of a snail shell.

Lateral: To the right or left of the midline.

Lateral line: A series of modified scales with pore-like openings to a sensory canal along the side of a fish.

Lirae: Fine lines or ridges (much finer than ribs) on mollusk shells.

Madreporite: An opening used to filter water into the water vascular system of echinoderms; wart-like structure on the aboral surface of a sea star.

Mandible: The lower jaw.

Mantle: In mollusks, the dorsal body wall covering the main body or visceral mass.

Maxilla (pl. maxillae): The upper jaw. In fish, maxillae are the hindmost bones of the upper jaw, preceded by the premaxillaries in front or below; also called maxillary bone (see p. 16).

Maxillaries: Pair of bones that form the back part of the lower jaw, often with teeth.

Merus: Fourth article or segment of an appendage, in crabs.

Ocellus (pl. ocelli): A round-to-oval eyelike pigment spot.

Operculum: In many gastropod mollusks, a horny plate that covers the opening of the shell when the animal is retracted; in higher fishes, the bony flap covering the gills.

Oral: In sea stars and brittle stars, the ventral side of the animal containing the mouth.

Osculum (pl. oscula): In sponges and some sea stars, a relatively large opening through which water flows out of the body mass.

Ossicles: In sea stars and brittle stars, the skeleton consists of multitudes of small calcareous plates (made of calcium carbonate) called ossicles that move with one another, forming flexible joints. In sea cucumbers the calcareous plates have degenerated and are buried in the fleshy body.

Pallial line: Line along which the shell is attached to the mantle, in bivalves.

Pallial sinus: Notch in the pallial line that marks the attachment of the muscle that retracts the siphon in bivalves.

Pandalid shrimp: Shrimp in the family Pandalidae. *Pandalus* and *Pandalopsis* are two shrimp genera in this family.

Papilla (pl. papillae): A small elevation or modified tube foot not used for locomotion, in sea cucumbers and other echinoderms.

Parapodia: One of the fleshy paired appendages of polychaete annelids that function in locomotion and breathing.

Paxilla (pl. paxillae): A pillar-like spine in some sea stars that sometimes has a flattened summit covered with spinules.

Pectoral fins: The front-most pair of ventral fins, in fish.

Pedicel (or pedicle): A small stalk or stalk-like support.

Pedicellaria (pl. pedicellariae): One of the tiny pincer-like structures of certain echinoderms.

Pelagic: Living in the open sea free from the bottom, usually near the surface.

Pelvic fins: Either of the pair of fins on the ventral surface of a fish.

Periostracum: The hard chitinous outermost covering of the shell of many mollusks.

Peristomium: The lip or edge of the aperture of a spiral shell; the membrane surrounding the mouth of an invertebrate.

Peritoneum: Membrane lining the body cavity.

Pinnate: Branched coral colonies in which the branching pattern is feather-like and in one plane.

Pleon: An abdominal plate of an isopod.

Pleopod: In shrimp and other crustaceans, paired or single abdominal appendages that may occur on the first five body segments; used for swimming and/or as a deposition site of fertilized eggs in females.

Pleotelson: A complex formed by the fusion of the telson with one or more pleons of an isopod.

Pleuron (pl. pleura): One of the paired lateral flaps on each side of the first five abdominal body segments in crustaceans.

Polyp: Any individual of a cnidarian colony regardless of anatomical structure; body cylindrical and an oral opening usually surrounded by tentacles.

Precaudal length: In skates, the distance from the tip of the snout to the posterior margin of the cloaca.

Preopercle: In fish, the bone in the operculum in front of the opercle and nearly parallel to it; its posterior edge is usually free and often has spines (see p. 16 and 27).

Proboscis: Any tubular process of the head or anterior part of the gut, usually used in feeding.

Process: A prominent or projecting part of an organism.

Proximal: Nearest to the body (opposite of distal).

Pseudobranch: Small, gill-like organ on the inner surface of the operculum of fishes.

Quadrate: In fish, paired triangular bones on which the mandible is hinged, connecting the lower jaw to the palatine and hyoid arches. In clams, a squarish shape.

R (major radius): A body dimension of a sea star measured from the center of the disc to the tip of the longest arm.

Rachis: The terminal or distal portion of a sea pen colony that bears the feeding polyps.

Radial rib: In gastropods, a ridge oriented along the length of a whorl so that it spirals around the shell with the whorl; also referred to a spiral rib.

Radial sculpture: In gastropods, a relief pattern from the apex outward to edge of shell; also referred to as spiral sculpture (see p. 146 and p. 185).

Radial: Developing symmetrically about a central point; echinoderms have a five-part radial symmetry.

Radiole: A spine on a sea urchin.

Reticulate: Marked with a network of lines.

Rhinophore: Paired sensory apparatus found on the head of some mollusks.

Rostrum: A bony projection or shelf on the snout of a fish; the forward projection of the carapace of a crab.

Scaphocerite: In many crustaceans, a flattened plate or scale attached to the second joint of the antenna.

Sculpture: Relief pattern on the surface of a shell.

Scute: A thickened bony scale bearing a keel or bony point in fish.

Scutum: In barnacles, one of a pair of plates that form part of the carapace.

Septa: The calcareous partitions of scleractinian coral chalices.

Serrated: Notched on the edge like a saw.

Seta (pl. setae): A stiff hair, bristle, or bristle-like part.

Sinistral: In snails, shell coils or turns from right to left or left-handed; opposite of dextral.

Sinus: A depression, curve, indentation, or any of various cavities or passages, as a hollow in a bone.

SL (standard length): The distance from the anterior tip of the head to the base of the caudal fin at the end of the vertebral column in fish; the posterior point can be found by flexing the caudal fin to produce a crease.

Spicule: A small needle-like structure of silicate or calcium that supports soft tissue.

Spider crab: Spider crab was a common name given to crabs of the genus *Chionoecetes*. Members of this genus have four pairs of walking legs like spiders. It's now commonly used to describe any crab with four pairs of walking legs.

Spinelet: A small, fixed spiny structure.

Spiral sculpture: See radial sculpture.

Spire: In snails, all the whorls above the body whorl.

Supraocular: Above the eye.

Suture: In snails, the suture divides one whorl from another.

Symphyseal knob: Projection or swelling below and in front of the symphysis of the lower jaw, in fish.

Tail length: In skates, the distance from the posterior margin of the cloaca to the tip of the tail.

Teeth, cardinal: In bivalves, the largest two or three ridges in the hinge.

Teeth, lateral: In bivalves, the smaller long, narrow ridges in the hinge below the cardinal teeth.

Telson: Rearmost segment of the body of some arthropods.

Test: Sea urchins and sand dollars have ossicles that are fused, forming a rigid skeletal shell known as the test.

TL (total length): The distance from the anterior-most tip of the snout to the posterior-most tip of the tail.

Tube feet: Numerous external, fluid-filled tubes of echinoderms, such as sea stars or sea urchins, serving as organs of locomotion, attachment, food handling, and respiration.

Tubercle: A small, rounded elevation on the body surface with a basal diameter equal to or greater than its height.

Umbilicus: In gastropods, a deep pit or opening at the base of the shell that is visible from below; the umbilicus leads into the columella, which is surrounded by the body whorl as the shell grows. The size of the umbilicus varies from closed in some genera, to rather wide in others.

Umbo: See beak.

Uniramous: Unbranched, referring to a crustacean appendage in which only the exopodite or endopodite of the biramous type of appendage persists.

Valve: A synonym for shell. It may refer to a snail shell, one of the two shells of a bivalve, or to one of the eight shell plates of a chiton.

Veil: In nudibranchs the veil is a membranous extension of the head.

Ventral: The undersurface, abdomen, or belly; opposite of dorsal.

Verrucose: Covered with small dermal warts.

Vomer: The median, sometimes toothed, bone at the back of the roof of the mouth, in fish.

Whorl: One of the turns of a gastropod spiral shell. In corals and hydroids, an arrangement of three or more branches or polyps radiating from a single node.

PHYLOGENIC SPECIES LIST

PHYLUM PORPHYRA

Class HEXACTINELLIDA (glass sponge)

Family Rossellidae
Vase sponge *Aulosaccus schulzei*

Class CALCAREA (calcareous sponge)

Family Grantiidae
Hairy vase sponge *Leucandra* sp.

Class DEMOSPONGIAE (variable skeleton sponges)

Family Geodiidae
Soccer ball sponge *Geodia mesotriaena*

Family Pachastrellidae
Fuzzy cratered sponge *Vulcanella* sp.

Family Tetillidae
Furry ball sponge *Craniella spinosa*
Baseball sponge *Craniella* sp.

Family Suberitidae
Hermit sponges *Suberites ficus*, *Suberites domuncula*

Family Polymastiidae
Nipple sponge *Polymastia* sp.

Family Mycalidae
Tree sponge *Mycale loveni*
Cotton ball sponge *Mycale lobata*

Family Acarnidae
Lattice sponge *Melonchela clathriata*

Family Coelosphaeridae
Spud sponge *Histodermella* sp.

Family Isodictyidae
Firm fingered sponge *Neoesperiopsis* sp.
Rough China hat sponge *Neoesperiopsis infundibula*

Family Myxillidae
Soft brown sponge *Myxilla* sp.
Scallop sponge *Myxilla incrustans*

PHYLUM CNIDARIA

Class HYDROZOA (hydroids)

Family Bonneviellidae
Champagne flute hydroid *Bonneviella* sp.

Family Aglaopheniidae
Ostrich-plume hydroid *Aglaophenia* sp.

Family Stylasteridae (hydrocorals)
Stylaster hydrocoral *Stylaster campylecus parageus*
Common hydrocoral *Stylaster campylecus campylecus*
Pink hydrocoral *Stylaster verrillii*
Capped hydrocoral *Crypthelia trophostega*
Rosy-undulate hydrocoral *Cyclohelia lamellata*
Rough hydrocoral *Errinopora zarhyncha*

Class SCYPHOZOA (jellyfish)

Family Pelagiidae
Sunrise jellyfish *Chrysaora melanaster*

Class ANTHOZOA (corals, sea fans, sea whip, sea pen, sea anemones)

Family Alcyoniidae
Red mushroom coral *Anthomastus* sp.
Gray mushroom coral *Anthomastus* sp.
Sea raspberry *Gersemia* sp.

Family Anthothelidae
Purple soft coral *Anthothela grandiflora*

Family Paragorgiidae
Bubblegum coral *Paragorgia arborea*

Family Acanthogorgiidae
Sea fan *Calcigorgia* sp.

Family Plexauridae
Purple sea fan *Muriceides nigra*
Kool's candelabrum gorgonian *Cryogorgia koolsae*
Pacific red sea fan *Swiftia* sp.

Family Primnoidae
Pinnate amphilaphis *Amphilaphis* sp.
Plumarella coral *Plumarella* sp.
Pacific red tree coral *Primnoa pacifica*
Utinomi's armored sea fan *Arthrogorgia utinomii*
Superb thouarella *Thouarella superba*
Giant fanellia *Fanellia compressa*

Family Pennatulidae (sea pen)
Orange sea pen *Ptilosarcus gurneyi*

Family Halipteridae (sea whip)
Giant sea whip *Halipteris willemoesi*

Family Caryophylliidae (cup coral)
Alaska cup coral *Caryophyllia alaskensis*

Family Actiniidae (anemones)
Mottled anemone *Urticina crassicornis*
Chevron-tentacled anemone *Cribrinopsis fernaldi*

Family Metridiidae (anemone)
Gigantic anemone (white plumed) *Metridium farcimen*

PHYLUM MOLLUSCA

Class POLYPLACOPHORA (chitons)

Family Acanthochitonidae
Giant Pacific (gumboot) chiton *Cryptochiton stelleri*

Family Mopaliidae
Black katy *Katharina tunicata*
White veiled chiton *Placiphorella pacifica*

Family Ischnochitonidae
Bering chiton *Lepidozona beringiana*

Class GASTROPODA (limpets, snails, whelks, nudibranchs)

Family Fissurellidae
Rough keyhole limpet *Diodora aspera*

Family Acmaeidae
Whitecap limpet *Acmaea mitra*

Family Calyptraeidae
Great slippersnail *Crepidula grandis*

Family Naticidae
Pale moonsnail *Euspira pallida*
Rusty moonsnail *Cryptonatica russa*
Aleutian moonsnail *Cryptonatica aleutica*
Arctic moonsnail *Cryptonatica affinis*

Family Capulidae
Two-keel hairysnail *Trichotropis bicarinata*

Family Ranellidae
Oregon triton *Fusitriton oregonensis*

Family Muricidae
Frilled dogwinkle *Nucella lamellosa*
Crown trophon *Boreotrophon coronatus*

Family Buccinidae
Sinuous whelk *Buccinum plectrum*
Ladder (silky) whelk *Buccinum scalariforme*
Polar whelk *Buccinum polaris*
Angular whelk *Buccinum angulosum*
Helmet whelk *Clinopegma magnum*
Thick-cord whelk *Beringius crebricostatus*
Kennicott's beringius *Beringius kennicottii*
"Golden" beringius *Beringius* sp.
Bering's beringius *Beringius behringi*
Northern neptune *Neptunea heros*
Lyre neptune *Neptunea lyrata*
Fat whelk *Neptunea ventricosa*
Pribilof whelk *Neptunea pribiloffensis*
Little neptune *Neptunea communis borealis*
"Vesteraalen" neptune *Neptunea* sp.
Large melon whelk *Pyrulofusus melonis*
Left-hand melon whelk *Pyrulofusus harpa*
Warped whelk *Pyrulofusus deformis*
Tulip whelk *Volutopsius middendorffii*
Fragile whelk *Volutopsius fragilis*
Frilled whelk *Volutopsius trophonius*
Volute whelk *Volutopsius castanea*
Threaded whelk *Lussivolutopsius filosus*
Arctic whelk *Colus kroyeri*
Oblique whelk *Colus hypolispus*
Shrew whelk *Colus hallii*

Family Volutidae
Ribbed volute *Arctomelon tamikoae*
Alaska volute *Arctomelon stearnsii*

Family Discodorididae
Pacific sea-lemon *Anisodoris nobilis*
"Spotted" diaulula *Diaulula* sp.

Family Tritoniidae
Giant orange tochui *Tochuina tetraquetra*
Tritonid nudibranch *Tritonia* sp.

Family Polyceratidae
Sea-clown triopha *Triopha catalinae*

Family Dendronotidae
Dall's dendronotid *Dendronotus dalli*
Red frond-aeolis *Dendronotus rufus*

Family Tethyidae
Lion nudibranch *Melibe leonina*

Family Dironidae
Golden dirona *Dirona aurantia*

Family Flabellinidae
Predaceous aeolis *Flabellina trophina*

Family Aeolidiidae
Shag-rug aeolis *Aeolidia papillosa*

Class BIVALVIA (mussels, oyster, jingle, scallops, cockles, clams)

Family Mytilidae
Northern horsemussel *Modiolus modiolus*
Black mussel *Musculus niger*
Discordant mussel *Musculus discors*
Foolish mussel *Mytilus trossulus*

Family Ostreidae
Pacific oyster *Crassostrea gigas*

Family Anomiidae
Alaska (green false) jingle *Pododesmus macroschisma*

Family Pectinidae
Reddish scallop *Chlamys rubida*
Weathervane scallop *Patinopectin caurinus*

Family Cardiidae
Hairy cockle *Clinocardium ciliatum*
Nuttall cockle *Clinocardium nuttallii*
Greenland smoothcockle *Serripes groenlandicus*
Oblique smoothcockle *Serripes notabilis*
Broad smoothcockle (broad cockle) *Serripes laperousii*

Family Veneridae
Washington butterclam *Saxidomus gigantea*
Pacific littleneck *Protothaca staminea*

Family Yoldiidae
Yoldia *Yoldia* sp.
Broad yoldia *Megayoldia thraciaeformis*

Family Thyasiridae
Disjunct cleft clam *Conchocele disjuncta*
Giant cleft clam *Conchocele bisecta*

Family Myidae
Elegant soft shell *Mya elegans*

Family Tellinidae
Alaska great-tellin *Tellina lutea*

Family Hiatellidae
Arctic roughmya *Panomya norvegica*
Arctic hiatella *Hiatella arctica*

Family Pholadidae
Pilsbry piddock *Zirfaea pilsbryi*

Family Pharidae
Alaska razor clam *Siliqua alta*
Pacific razor clam *Siliqua patula*

Family Mactridae
Arctic surfclam *Mactromeris polynyma*
Fat gaper clam *Tresus capax*

Family Astartidae
Arctic astarte *Astarte arctica*
Elliptical astarte *Astarte elliptica*

Class CEPHALOPODA (squids, octopus)

Family Gonatidae
Magister armhook squid *Berryteuthis magister*

Family Sepiolidae
North Pacific bobtail squid *Rossia pacifica*

Family Octopodidae
Pacific giant octopus *Enteroctopus dofleini*

PHYLUM ECTOPROCTA (BRYOZOA)

Class GYMNOLAEMATA (bryozoans)

Family Calpensiidae
Jointed bryozoan *Microporina borealis*

Family Bugulidae
Fan bryozoan *Dendrobeania* sp.
Spiral bryozoan *Bugula californica*

Family Alcyonidiidae
Smooth leather bryozoan *Alcyonidium pedunculatum*
Medusa bryozoan *Alcyonidium* sp.

Family Flustrellidridae
Spiny leather bryozoan *Flustrellidra corniculata*

PHYLUM BRACHIOPODA

Class ARTICULATA (brachiopods)

Family Hemithyrididae
Black lamp shell *Hemithiris psittacea*

Family Laqueidae
Transverse lamp shell *Terebratalia transversa*

PHYLUM ANNELIDA

Class POLYCHAETA (worms)

Family Aphroditidae
Sea mouse *Aphrodita negligens*

Family Polynoidae
Depressed scale worm *Eunoe depressa*
Giant scale worm *Eunoe nodosa*
Red-banded commensal worm *Arctonoe vittata*

Family Sabellidae
Northern feather duster worm *Eudistylia vancouveri*

Family Glyceridae
Northern iridescent worm *Hemipodus borealis*

Family Serpulidae
Calcareous tube worm *Serpula vermicularis*

Class CLITELLATA (leech)

Family Piscicolidae
Striped sea leech *Notostomum cyclostomum*

PHYLUM ARTHROPODA

Class PYCNOGONIDA (sea spider)

Family Colossendeidae
Giant sea spider *Colossendeis* sp.

Class MAXILLOPODA (barnacles)

Family Scalpellidae
Eared barnacle *Tarasovium cornutum*

Family Lepadidae
Pelagic gooseneck barnacle *Lepas* sp.

Family Balanidae
Beaked barnacle *Balanus rostratus*

Family Archaeobalanidae
Giant barnacle *Chirona evermanni*
Crab barnacle *Hesperibalanus hesperius*

Class MALACOSTRACA (isopod, amphipod, shrimps, crabs)

Order Isopoda
Family Idoteidae
Gravel isopod *Synidotea* sp.

Family Aegidae
Sea cockroach *Rocinela angustata*

Family Arcturidae
Spiky arcturid *Arcturus* sp.
Spiny arcturid *Arcturus* sp.

Order Amphipoda
Family Caprellidae
Skeleton shrimp *Caprella* sp.

Order Decapoda
Family Pandalidae
Sidestripe shrimp *Pandalopsis dispar*
Aleutian bigeye *Pandalopsis aleutica*
Spot shrimp (prawn) *Pandalus platyceros*
Coonstripe shrimp *Pandalus hypsinotus*
Dock shrimp *Pandalus danae*
Alaskan pink shrimp *Pandalus eous*
Humpy shrimp *Pandalus goniurus*
Yellowleg pandalid *Pandalus tridens*

Family Hippolytidae
Spiny lebbeid *Lebbeus groenlandicus*

Family Crangonidae
Arctic argid *Argis dentate*
Sculptured shrimp *Sclerocrangon boreas*

Order Decapoda
Family Lithodidae
Puget Sound king crab *Lopholithodes mandtii*
Brown box crab *Lopholithodes foraminatus*
Rhinoceros crab *Rhinolithodes wosnessenskii*
Heart crab *Phyllolithodes papillosus*
Spiny lithode crab *Acantholithodes hispidus*
Soft crab *Hapalogaster grebnitzkii*
Hairy crab *Hapalogaster mertensii*
Butterfly crab *Cryptolithodes typicus*
Red king crab *Paralithodes camtschaticus*
Blue king crab *Paralithodes platypus*
Spiny king crab *Paralithodes brevipes*
Golden king crab *Lithodes aequispinus*
Scarlet king crab *Lithodes couesi*
Scaled crab *Placetron wosnessenskii*
Verrill's spider crab *Paralomis verrilli*
Spiny paralomis *Paralomis multispina*
Spiny spider crab *Paralomis makarovi*

Family Galatheidae
Pinchbug *Munida quadrispina*

Family Oregoniidae
Triangle Tanner crab *Chionoecetes angulatus*
Grooved Tanner crab *Chionoecetes tanneri*
Tanner crab *Chionoecetes bairdi*
Snow crab *Chionoecetes opilio*
Hybrid Tanner crab *Chionoecetes bairdi × opilio*
Graceful decorator crab *Oregonia gracilis*
Pacific lyre crab *Hyas lyratus*
Arctic lyre crab *Hyas coarctatus*

Family Epialtidae
Graceful kelp crab *Pugettia gracilis*

Family Pisidae
Longhorn decorator crab *Chorilia longipes*

Family Cheiragonidae
Hair crab *Erimacrus isenbeckii*
Helmet crab *Telmessus cheiragonus*

Family Cancridae
Dungeness crab *Cancer magister*
Pygmy rock crab *Cancer oregonensis*

Family Pinnotheridae
Gaper pea crab *Pinnixa littoralis*

Family Paguridae
Splendid hermit *Labidochirus splendescens*
Widehand hermit *Elassochirus tenuimanus*
Purple hermit *Elassochirus cavimanus*
Pacific red hermit *Elassochirus gilli*
Aleutian hermit *Pagurus aleuticus*
Alaskan hermit *Pagurus ochotensis*
Longfinger hermit *Pagurus rathbuni*
"Dirty harry" hermit *Pagurus capillatus*
Fuzzy hermit *Pagurus trigonocheirus*
Whiteknee hermit *Pagurus dalli*
Knobbyhand hermit *Pagurus confragosus*

PHYLUM ECHINODERMATA

Class CRINOIDEA (feather star)

Family Antedonidae
Feather star *Florometra* sp.

Class ASTEROIDEA (sea stars)

Family Astropectinidae
Arctic sand star *Leptychaster arcticus*

Family Goniopectinidae
Common mud star *Ctenodiscus crispatus*

Family Benthopectinidae
Fragile star *Cheiraster dawsoni*

Family Goniasteridae
Orange cookie star *Ceramaster patagonicus*
Arctic cookie star *Ceramaster arcticus*
Stellate cookie star *Ceramaster stellatus*
Red cookie star *Ceramaster japonicus*
Grainy white cookie star *Cladaster validus*
Kurile spiny star *Hippasteria kurilensis*
Pale spiny star *Hippasteria* sp.
Northern scarlet star *Pseudarchaster parelii*
Alaskan scarlet star *Pseudarchaster parelii alascensis*

Family Poraniidae
Leather star *Dermasterias imbricata*

Family Solasteridae
Striped sun star *Solaster stimpsoni*
Remarkable sun star *Solaster* sp.
Beautiful sun star *Solaster* sp.
Serpent sun star *Solaster* sp.

Rose star *Crossaster papposus*
Crested star *Lophaster furcilliger*

Family Pterasteridae
Tessellated slime star *Pteraster tesselatus*
Northern slime star *Pteraster* sp.
Obscure cushion star *Pteraster obscurus*
Eight-rayed cushion star *Pteraster octaster*
Wrinkled cushion star *Pteraster militaris*
Pincushion star *Diplopteraster multipes*

Family Echinasteridae
Blood star *Henricia* cf. *leviuscula*
Spiny henricia *Henricia multispina*
Tumid henricia *Henricia tumida*
Bering henricia *Henricia beringiania*
Fuzzy henricia *Henricia* sp.
Purple toothed henricia *Odontohenricia* sp.
Pale toothed henricia *Odontohenricia* sp.

Family Asteriidae
Purple-orange star *Asterias amurensis*
Giant star *Evasterias echinosoma*
Mottled star *Evasterias troschelii*
Arctic star *Leptasterias arctica*
Aleutian star *Leptasterias hylodes*
Knobby six-rayed star *Leptasterias polaris*
Giant Aleutian six-rayed star *Leptasterias truculenta*
Sunflower star *Pycnopodia helianthoides*
Redbanded star *Orthasterias koehleri*
Blackspined star *Lethasterias nanimensis*
Long-rayed star *Stylasterias forreri*
Ochre star *Pisaster ochraceus*

Class OPHIUROIDEA (basket stars, serpent stars, brittle stars)

Family Gorgonocephalidae
Basket star *Gorgonocephalus eucnemis*

Family Asteronychidae
Dwarf smooth serpent star *Astronebris tatafilius*
Giant serpent star *Asteronyx loveni*

Family Gorgonocephalidae
White serpent star *Astrochele laevis*

Family Ophiactidae
Daisy brittle star *Ophiopholis aculeata kennerlyi*
Ubiquitous brittle star *Ophiopholis aculeata aculeata*

Famly Ophiacanthidae
Spiny-brown brittle star *Ophiacantha cataleimmoidus*
Nine-armed brittle star *Ophiacantha enneactis*
Nodulose brittle star *Ophiolebes* sp.

Family Ophiuridae
Notched brittle star *Ophiura sarsi*
Great armored brittle star *Stegophiura ponderosa*

Class ECHINOIDEA (sea urchins, heart urchin, sand dollar)

Family Strongylocentrotidae
Aleutian urchin *Strongylocentrotus polyacanthus*
Green urchin *Strongylocentrotus droebachiensis*
Red urchin *Strongylocentrotus franciscanus*
Fragile (pink) urchin *Strongylocentrotus fragilis*

Family Schizasteridae
Heart urchin *Brisaster latifrons*

Family Echinarachniidae
Northern (Parma) sand dollar *Echinarachnius parma*

Class HOLOTHUROIDEA (sea cucumbers)

Family Stichopodidae
Giant California cucumber *Parastichopus californicus*

Family Synallactidae
Northern papillate (slender) cucumber *Bathyplotes* sp.

Family Psolidae
Red-orange armored cucumber *Psolus japonicus*
White-scaled cucumber *Psolus squamatus*
Arctic armored cucumber *Psolus fabricii*

Family Cucumariidae
Sea football *Cucumaria fallax*

Family Phyllophoridae
Crescent cucumber *Pentamera lissoplaca*

Family Molpadiidae
Sweet sea potato *Molpadia intermedia*

PHYLUM CHORDATA

Class ASCIDIACEA (tunicates)

Family Pyuridae
Sea onion *Boltenia ovifera*
Bristly tunicate *Halocynthia hilgendorfi igaboja*
Sea peach *Halocynthia aurantium*

Family Styelidae
Sea potato *Styela rustica*

Class ELASMOBRANCHII (cartilaginous fishes)

Order Squaliformes (sharks)
Family Somniosidae
Pacific sleeper shark *Somniosus pacificus*

Family Squalidae
Spiny dogfish *Squalus acanthias*

Order Rajiformes (skates)
Family Rajidae
Longnose skate *Raja rhina*
Big skate *Raja binoculata*

Family Arhynchobatidae
Alaska skate *Bathyraja parmifera*
"Leopard" skate *Bathyraja* sp. cf. *parmifera*
Mud skate *Bathyraja taranetzi*
Aleutian skate *Bathyraja aleutica*
Bering skate *Bathyraja interrupta*

Commander skate *Bathyraja lindbergi*
Whiteblotched skate *Bathyraja maculata*

Superclass Agnatha (jawless fishes) Class CEPHALASPIDOMORPHI (lamprey)

Order Petromyzontiformes

Family Petromyzontidae

Arctic lamprey *Lethenteron camtschaticum*

Superclass Osteichthyes (bony fishes) CLASS ACTINOPTERYGII (ray-finned fishes)

Order Salmoniformes

Family Salmonidae (salmon, trouts)

Rainbow trout *Oncorhynchus mykiss*
Cutthroat trout *Oncorhynchus clarkii*
Dolly Varden *Salvelinus malma*
Pink (humpback) salmon *Oncorhynchus gorbuscha*
Coho (silver) salmon *Oncorhynchus kisutch*
Chinook (king) salmon *Oncorhynchus tshawytscha*
Chum (dog) salmon *Oncorhynchus keta*
Sockeye (red) salmon *Oncorhynchus nerka*

Order Clupeiformes

Family Clupeidae (herring)

Pacific herring *Clupea pallasii*

Order Osmeriformes

Family Osmeridae (smelts)

Rainbow smelt *Osmerus mordax*
Capelin *Mallotus villosus*
Eulachon *Thaleichthys pacificus*

Order Gadiformes

Family Gadidae (cods)

Pacific (gray) cod *Gadus macrocephalus*
Pacific tomcod *Microgadus proximus*
Saffron cod *Eleginus gracilis*
Walleye pollock *Theragra chalcogramma*

Order Scorpaeniformes

Family Scorpaenidae (rockfishes)

Shortspine thornyhead *Sebastolobus alascanus*
Yelloweye rockfish *Sebastes ruberrimus*
Shortraker rockfish *Sebastes borealis*
Rougheye rockfish *Sebastes aleutianus*
Blackspotted rockfish *Sebastes melanostictus*
Pacific ocean perch *Sebastes alutus*
Harlequin rockfish *Sebastes variegatus*
Sharpchin rockfish *Sebastes zacentrus*
Northern rockfish *Sebastes polyspinis*
Canary rockfish *Sebastes pinniger*
Tiger rockfish *Sebastes nigrocinctus*
Redbanded rockfish *Sebastes babcocki*
China rockfish *Sebastes nebulosus*
Black rockfish *Sebastes melanops*
Dark rockfish *Sebastes ciliatus*
Dusky rockfish *Sebastes variabilis*
Silvergray rockfish *Sebastes brevispinis*
Yellowtail rockfish *Sebastes flavidus*
Quillback rockfish *Sebastes maliger*

Family Anoplopomatidae (sablefish)

Sablefish (blackcod) *Anoplopoma fimbria*

Family Hexagrammidae (greenlings)

Atka mackerel *Pleurogrammus monopterygius*
Lingcod *Ophiodon elongatus*
Kelp greenling *Hexagrammos decagrammus*
Rock greenling *Hexagrammos lagocephalus*
Masked greenling *Hexagrammos octogrammus*

Family Rhamphocottidae (grunt sculpins)

Grunt sculpin *Rhamphocottus richardsonii*

Family Cottidae (sculpins)

Roughspine sculpin *Triglops macellus*
Scissortail sculpin *Triglops forficatus*
Red Irish lord *Hemilepidotus hemilepidotus*
Yellow Irish lord *Hemilepidotus jordani*
Threaded sculpin *Gymnocanthus pistilliger*
Armorhead sculpin *Gymnocanthus galeatus*
Leister sculpin *Enophrys lucasi*
Brightbelly sculpin *Microcottus sellaris*
Shorthorn sculpin *Myoxocephalus scorpius*
Plain sculpin *Myoxocephalus jaok*
Great sculpin *Myoxocephalus polyacanthocephalus*

Family Hemitripteridae (sailfin sculpins)

Bigmouth sculpin *Hemitripterus bolini*
Crested sculpin *Blepsias bilobus*
Silverspotted sculpin *Blepsias cirrhosus*
Sailfin sculpin *Nautichthys oculofasciatus*

Family Psychrolutidae (fathead sculpins)

Spinyhead sculpin *Dasycottus setiger*
Darkfin sculpin *Malacocottus zonurus*

Family Agonidae (sea poachers)

Sawback poacher *Leptagonus frenatus*
Fourhorn poacher *Hypsagonus quadricornis*
Sturgeon poacher *Podothecus accipenserinus*
Blackfin poacher *Bathyagonus nigripinnis*
Alligatorfish *Aspidophoroides monopterygius*

Family Cyclopteridae (lumpsucker)

Smooth lumpsucker *Aptocyclus ventricosus*

Family Liparidae (snailfishes)

Variegated snailfish *Liparis gibbus*
Salmon snailfish *Careproctus rastrinus*
Spectral snailfish *Careproctus phasma*
Blotched snailfish *Crystallichthys cyclospilus*
Combed snailfish *Allocareproctus kallaion*
Cherry snailfish *Allocareproctus jordani*

Order Perciformes (perch-like fishes)

Family Pentacerotidae (armorhead)

North Pacific armorhead *Pseudopentaceros wheeleri*

Family Bathymasteridae (ronquils)

Searcher *Bathymaster signatus*
Alaskan ronquil *Bathymaster caeruleofasciatus*

Family Zoarcidae (eelpouts)
Shortfin eelpout *Lycodes brevipes*
Wattled eelpout *Lycodes palearis*
Polar eelpout *Lycodes turneri*
Alaska eelpout *Bothrocara pusillum*

Family Stichaeidae (pricklebacks)
Decorated warbonnet *Chirolophis decoratus*
Longsnout prickleback *Lumpenella longirostris*
Snake prickleback *Lumpenus sagitta*
Whitebarred prickleback *Poroclinus rothrocki*
Blackline prickleback *Acantholumpenus mackayi*
Stout eelblenny *Anisarchus medius*

Family Cryptacanthodidae (wrymouths)
Giant wrymouth *Cryptacanthodes giganteus*
Dwarf wrymouth *Cryptacanthodes aleutensis*

Family Pholidae (gunnel)
Crescent gunnel *Pholis laeta*

Family Anarhichadidae (wolffish)
Bering wolffish *Anarhichas orientalis*

Family Zaproridae (prowfish)
Prowfish *Zaprora silenus*

Family Trichodontidae (sandfish)
Pacific sandfish *Trichodon trichodon*

Family Ammodytidae (sand lance)
Pacific sand lance *Ammodytes hexapterus*

Order Pleuronectiformes (flatfishes, flounders, soles)

Family Pleuronectidae (righteye flounders)
Pacific halibut *Hippoglossus stenolepis*
Greenland turbot *Reinhardtius hippoglossoides*
Arrowtooth flounder *Atheresthes stomias*
Kamchatka flounder *Atheresthes evermanni*
Flathead sole *Hippoglossoides elassodon*
Butter sole *Isopsetta isolepis*
Rex sole *Glyptocephalus zachirus*
Southern rock sole *Lepidopsetta bilineata*
Northern rock sole *Lepidopsetta polyxystra*
Dover sole *Microstomus pacificus*
Yellowfin sole *Limanda aspera*
English sole *Parophrys vetulus*
Starry flounder *Platichthys stellatus*
Alaska plaice *Pleuronectes quadrituberculatus*

INDEX TO COMMON AND SCIENTIFIC NAMES

A

Acantholithodes hispidus 222
Acantholumpenus mackayi 82
Acmaea mitra 144
Aeolidia papillosa 178
Aglaophenia sp. 114
Alaska cup coral 130
Alaska eelpout 78
Alaska great-tellin 194
Alaska (green false) jingle 182
Alaska plaice 102
Alaska razor clam 196
Alaska skate 10
Alaska volute 168
Alaskan hermit 246
Alaskan pink shrimp 214
Alaskan ronquil 74
Alaskan scarlet star 262
Alcyonidium pedunculatum 252
Alcyonidium sp. 252
Aleutian bigeye 212
Aleutian hermit 244
Aleutian moonsnail 148
Aleutian skate 12
Aleutian star 278
Aleutian urchin 294
alligatorfish 66
Allocareproctus jordani 72
Allocareproctus kallaion 70
Allocentrotus fragilis (see *Strongylocentrotus fragilis*)
Ammodytes hexapterus 90
Amphilaphis sp. 126
Amphiophiura ponderosa (see *Stegophiura ponderosa*)
amphipod 210
Anarhichas orientalis 86
angular whelk 154
Anisarchus medius 82
Anisodoris nobilis 172
Anoplopoma fimbria 44
Anthomastus sp. 120
Anthothela grandiflora 122
Aphrodita negligens 136
Aptocyclus ventricosus 66
arctic argid 218
arctic armored cucumber 300
arctic astarte 198
arctic cookie star 258
arctic hiatella 194
arctic lamprey 4
arctic lyre crab 236
arctic moonsnail 148
arctic roughmya 194
arctic sand star 256
arctic star 278
arctic surfclam 198
arctic whelk 166
Arctomelon stearnsii 168
Arctomelon tamikoae 168
Arctonoe vittata 138
Arcturus sp. 208, 210
Argis dentata 218
armorhead 73
armorhead sculpin 54
arrowtooth flounder 92
Arthrogorgia utinomii 128
Aspidophoroides monopterygius 66
Astarte arctica 198
Astarte elliptica 200
Asterias amurensis 276
Asteronyx loveni 287
Astrochele laevis 287
Astronebris tatafilius 286
Atheresthes evermanni 92
Atheresthes stomias 92
Atka mackerel 44
Aulosaccus schulzei 103

B

Balanus rostratus 206
barnacles 204
baseball sponge 106
basket star 284
Bathyagonus nigripinnis 64
Bathymaster caeruleofasciatus 74
Bathymaster signatus 74
Bathyplotes sp. 298
Bathyraja aleutica 12
Bathyraja interrupta 12
Bathyraja lindbergi 14
Bathyraja maculata 14
Bathyraja parmifera 10
Bathyraja sp. cf. *parmifera* 10
Bathyraja taranetzi 10
beaked barnacle 206
beautiful sun star 266
Bering chiton 142
Bering henricia 274
Bering skate 12
Bering wolffish 86
Beringius behringi 156

Beringius crebricostatus 154
beringius eggs 170
Beringius kennicottii 156
Beringius sp. 156
Beringius sp. eggs 170
Bering's beringius 156
Berryteuthis magister 202
big skate 8
bigmouth sculpin 58
black Katy 140
black lamp shell 200
black mussel 180
black rockfish 40
blackcod 44
blackfin poacher 64
blackline prickleback 82
blackspined star 282
blackspotted rockfish 32
Blepsias bilobus 59
Blepsias cirrhosus 60
blood star 272
blotched snailfish 70
blue king crab 226
Boltenia ovifera 302
Bonneviella sp. 114
Boreotrophon coronatus 150
Bothrocara pusillum 78
brachiopods 200
brightbelly sculpin 54
Brisaster latifrons 296
bristly tunicate 304
brittle stars 286
broad smoothcockle (broad cockle) 188
broad yoldia 190
brown box crab 220
bryozoans 250
bubblegum coral 122
Buccinum angulosum 154
buccinum eggs 170
Buccinum plectrum 152
Buccinum polaris 152
Buccinum scalariforme 152
Buccinum sp. eggs 170
Bugula californica 250
butter sole 94
butterfly crab 224

C

calcareous sponge 104
calcareous tube worm 140
Calcigorgia sp. 122
canary rockfish 36
Cancer magister 240
Cancer oregonensis 240
capelin 24
capped hydrocoral 118
Caprella sp. 210
Careproctus phasma 70
Careproctus rastrinus 69
Caryophyllia alaskensis 130
Ceramaster arcticus 258
Ceramaster japonicus 260
Ceramaster patagonicus 258
Ceramaster stellatus 258
champagne flute hydroid 114
Cheiraster dawsoni 256
cherry snailfish 72
chevron-tentacled anemone 132
China rockfish 38
chinook (king) salmon 20
Chionoecetes angulatus 232
Chionoecetes bairdi 234
Chionoecetes bairdi × opilio 234
Chionoecetes opilio 234
Chionoecetes tanneri 232
Chirolophis decoratus 80
Chirona evermanni 206
chitons 140
Chlamys rubida 182
Chorilia longipes 238
Chrysaora melanaster 134
chum (dog) salmon 20
Cladaster validus 260
clams 188
Clinocardium ciliatum 186
Clinocardium nuttallii 186
Clinopegma magnum 154
Clupea pallasii 22
cockles 186
cods 25
coho (silver) salmon 18
Colossendeis sp. 204
Colus hallii 168
Colus hypolispus 168
Colus kroyeri 166
combed snailfish 70
Commander skate 14
common hydrocoral 116
common mud star 256
Conchocele bisecta 192
Conchocele disjuncta 192
coonstripe shrimp 214
corals 116
cotton ball sponge 110
crab barnacle 206
crabs 220
Craniella sp. 106
Craniella spinosa 106
Crassostrea gigas 182
Crepidula grandis 144
crescent cucumber 302
crescent gunnel 86
crested sculpin 59
crested star 268
Cribrinopsis fernaldi 132

Crossaster papposus 266
crown trophon 150
Cryogorgia koolsae 124
Cryptacanthodes aleutensis 84
Cryptacanthodes giganteus 84
Crypthelia trophostega 118
Cryptochiton stelleri 142
Cryptolithodes typicus 224
Cryptonatica affinis 148
Cryptonatica aleutica 148
Cryptonatica russa 147
Crystallichthys cyclospilus 70
Ctenodiscus crispatus 256
Cucumaria fallax 300
cup coral 130
cutthroat trout 16
Cyclohelia lamellata 118

D

daisy brittle star 288
Dall's dendronotid 174
dark rockfish 40
darkfin sculpin 62
Dasycottus setiger 60
decorated warbonnet 80
Dendrobeania sp. 250
Dendronotus dalli 174
Dendronotus rufus 176
depressed scale worm 136
Dermasterias imbricata 264
Diaulula sp. 172
Diodora aspera 144
Diplopteraster multipes 270
Dirona aurantia 176
"dirty harry" hermit 246
discordant mussel 180
disjunct cleft clam 192
dock shrimp 214
dog salmon 20
Dolly Varden 18
Dover sole 98
Dungeness crab 240
dusky rockfish 40
dwarf smooth serpent star 286
dwarf wrymouth 84

E

eared barnacle 204
Echinarachnius parma 296
eelpouts 76
eight-rayed cushion star 270
Elassochirus cavimanus 244
Elassochirus gilli 244
Elassochirus tenuimanus 243
elegant soft shell 192
Eleginus gracilis 26
elliptical astarte 200
English sole 100
Enophrys lucasi 54
Enteroctopus dofleini 202
Erimacrus isenbeckii 238
Errinopora zarhyncha 118
Eudistylia vancouveri 138
eulachon 24
Eunoe depressa 136
Eunoe nodosa 136
Euspira pallida 147
Evasterias echinosoma 276
Evasterias troschelii 278

F

fan bryozoan 250
Fanellia compressa 128
fat gaper clam 198
fat neptune snail eggs 170
fat whelk 158
feather star 254
firm fingered sponge 112
Flabellina trophina 178
flathead sole 94
Florometra sp. 254
Flustrellidra corniculata 252
foolish mussel 180
fourhorn poacher 64
fragile (pink) urchin 296
fragile star 256
fragile whelk 164
frilled dogwinkle 150
frilled whelk 164
furry ball sponge 106
Fusitriton oregonensis 150
Fusitriton oregonensis eggs 170
fuzzy cratered sponge 106
fuzzy henricia 274
fuzzy hermit 248

G

Gadus macrocephalus 25
gaper pea crab 240
Geodia mesotriaena 104
Gersemia sp. 120
giant Aleutian six-rayed star 280
giant barnacle 206
giant California cucumber 298
giant cleft clam 192
giant fanellia 128
giant orange tochui 172
giant Pacific (gumboot) chiton 142
giant scale worm 136
giant sea spider 204
giant sea whip 130
giant serpent star 287
giant star 276
giant wrymouth 84

gigantic anemone 134
glass sponge 103
Glyptocephalus zachirus 96
"golden" beringius 156
golden dirona 176
golden king crab 228
Gorgonocephalus eucnemis 284
graceful decorator crab 234
graceful kelp crab 236
grainy white cookie star 260
gravel isopod 208
gray mushroom coral 120
great armored brittle star 292
great sculpin 56
great slippersnail 144
green urchin 294
Greenland smoothcockle 186
Greenland turbot 92
greenlings 44
grooved Tanner crab 232
grunt sculpin 49
gunnel 86
Gymnocanthus galeatus 54
Gymnocanthus pistilliger 53

H

hair crab 238
hairy cockle 186
hairy crab 224
hairy vase sponge 104
Halipteris willemoesi 130
Halocynthia aurantium 304
Halocynthia hilgendorfi igaboja 304
Hapalogaster grebnitzkii 224
Hapalogaster mertensii 224
harlequin rockfish 34
heart crab 222
heart urchin 296
helmet crab 238
helmet whelk 154
Hemilepidotus hemilepidotus 50
Hemilepidotus jordani 52
Hemipodus borealis 138
Hemithiris psittacea 200
Hemitripterus bolini 58
Henricia beringiania 274
Henricia cf. *leviuscula* 272
Henricia multispina 272
Henricia sp. 274
Henricia tumida 272
hermit crabs 242
hermit sponges 108
herring 22
Hesperibalanus hesperius 206
Hexagrammos decagrammus 46
Hexagrammos lagocephalus 46
Hexagrammos octogrammus 48
Hiatella arctica 194
Hippasteria kurilensis 260
Hippasteria sp. 262
Hippoglossoides elassodon 94
Hippoglossus stenolepis 90
Histodermella sp. 110
humpback salmon 18
humpy shrimp 216
Hyas coarctatus 236
Hyas lyratus 236
hybrid Tanner crab 234
hydrocorals 116
hydroids 114
Hypsagonus quadricornis 64

I

isopods 208
Isopsetta isolepis 94

J

jellyfish 134
jingle 182
jointed bryozoan 250

K

Kamchatka flounder 92
Katharina tunicata 140
kelp greenling 46
Kennicott's beringius 156
king salmon 20
knobby six-rayed star 280
knobbyhand hermit 248
Kool's candelabrum gorgonian 124
Kurile spiny star 260

L

Labidochirus splendescens 243
ladder (silky) whelk 152
lamprey 4
large melon whelk 162
lattice sponge 110
leather star 264
Lebbeus groenlandicus 216
leech 140
left-hand melon whelk 162
Leister sculpin 54
"leopard" skate 10
Lepas sp. 204
Lepidopsetta bilineata 96
Lepidopsetta polyxystra 98
Lepidozona beringiana 142
Leptagonus frenatus 63
Leptasterias arctica 278
Leptasterias hylodes 278
Leptasterias polaris 280
Leptasterias truculenta 280
Leptychaster arcticus 256

Lethasterias nanimensis 282
Lethenteron camtschaticum 4
Leucandra sp. 104
Limanda aspera 100
limpets 144
lingcod 44
lion nudibranch 176
Liparis gibbus 68
Lithodes aequispinus 228
Lithodes couesi 228
little neptune 160
long-rayed star 282
longfinger hermit 246
longhorn decorator crab 238
longnose skate 8
longsnout prickleback 80
Lophaster furcilliger 268
Lopholithodes foraminatus 220
Lopholithodes mandtii 220
Lumpenella longirostris 80
Lumpenus sagitta 80
lumpsucker 66
Lussivolutopsius filosus 166
Lycodes brevipes 76
Lycodes palearis 76
Lycodes turneri 78
lyre neptune 158
lyre neptune eggs 170

M

Mactromeris polynyma 198
magister armhook squid 202
Malacocottus zonurus 62
Mallotus villosus 24
masked greenling 48
medusa bryozoan 252
Megayoldia thraciaeformis 190
Melibe leonina 176
Melonchela clathriata 110
Metridium farcimen 134
Microcottus sellaris 54
Microgadus proximus 25
Microporina borealis 250
Microstomus pacificus 98
Modiolus modiolus 178
Molpadia intermedia 302
moonsnails 147
mottled anemone 132
mottled star 278
mud skate 10
Munida quadrispina 232
Muriceides nigra 124
Musculus discors 180
Musculus niger 180
mussels 178
Mya elegans 192
Mycale lobata 110
Mycale loveni 108
Myoxocephalus jaok 56
Myoxocephalus polyacanthocephalus 56
Myoxocephalus scorpius 56
Mytilus trossulus 180
Myxilla incrustans 114
Myxilla sp. 112

N

Nautichthys oculofasciatus 60
Neoesperiopsis infundibula 112
Neoesperiopsis sp. 112
Neptunea communis borealis 160
Neptunea heros 158
Neptunea heros eggs 170
Neptunea lyrata 158
Neptunea lyrata eggs 170
Neptunea pribiloffensis 160
Neptunea sp. 160
Neptunea ventricosa 158
Neptunea ventricosa eggs 170
nine-armed brittle star 290
nipple sponge 108
nodulose brittle star 291
North Pacific armorhead 73
North Pacific bobtail squid 202
northern feather duster worm 138
northern horsemussel 178
northern iridescent worm 138
northern neptune 158
northern neptune eggs 170
northern papillate (slender) cucumber 298
northern (Parma) sand dollar 296
northern rock sole 98
northern rockfish 36
northern scarlet star 262
northern slime star 268
notched brittle star 292
Notostomum cyclostomum 140
Nucella lamellosa 150
nudibranchs 172
Nuttall cockle 186

O

oblique smoothcockle 188
oblique whelk 168
obscure cushion star 270
ochre star 284
octocorals (soft corals) 120
octopus 202
Odontohenricia sp. 274
Oncorhynchus clarkii 16
Oncorhynchus gorbuscha 18
Oncorhynchus keta 20
Oncorhynchus kisutch 18
Oncorhynchus mykiss 16
Oncorhynchus nerka 20

Oncorhynchus tshawytscha 20
Ophiacantha cataleimmoidus 288
Ophiacantha enneactis 290
Ophiodon elongatus 44
Ophiolebes sp. 291
Ophiopholis aculeata aculeata 288
Ophiopholis aculeata kennerlyi 288
Ophiopthalmus cataleimmoidus (see *Ophiacantha cataleimmoidus*) 288
Ophiura sarsi 292
orange cookie star 258
orange sea pen 130
Oregon triton 150
Oregon triton eggs 170
Oregonia gracilis 234
Orthasterias koehleri 282
Osmerus mordax 22
ostrich-plume hydroid 114
oyster 182

P

Pacific giant octopus 202
Pacific (gray) cod 25
Pacific halibut 90
Pacific herring 22
Pacific littleneck 190
Pacific lyre crab 236
Pacific ocean perch 32
Pacific oyster 182
Pacific razor clam 196
Pacific red hermit 244
Pacific red sea fan 124
Pacific red tree coral 126
Pacific sand lance 90
Pacific sandfish 88
Pacific sea-lemon 172
Pacific sleeper shark 4
Pacific tomcod 25
Pagurus aleuticus 244
Pagurus capillatus 246
Pagurus confragosus 248
Pagurus dalli 248
Pagurus ochotensis 246
Pagurus rathbuni 246
Pagurus trigonocheirus 248
pale moonsnail 147
pale spiny star 262
pale toothed henricia 276
Pandalopsis aleutica 212
Pandalopsis dispar 212
Pandalus danae 214
Pandalus eous 214
Pandalus goniurus 216
Pandalus hypsinotus 214
Pandalus platyceros 212
Pandalus tridens 216
Panomya norvegica 194
Paragorgia arborea 122
Paralithodes brevipes 226
Paralithodes camtschaticus 226
Paralithodes platypus 226
Paralomis makarovi 230
Paralomis multispina 230
Paralomis verrilli 230
Parastichopus californicus 298
Parophrys vetulus 100
Patinopectin caurinus 184
pelagic gooseneck barnacle 204
Pentamera lissoplaca 302
Pholis laeta 86
Phyllolithodes papillosus 222
Pilsbry piddock 196
pinchbug 232
pincushion star 270
pink (humpback) salmon 18
pink hydrocoral 116
pinnate amphilaphis 126
Pinnixa littoralis 240
Pisaster ochraceus 284
Placetron wosnessenskii 228
Placiphorella pacifica 142
plain sculpin 56
Platichthys stellatus 100
Pleurogrammus monopterygius 44
Pleuronectes quadrituberculatus 102
plumarella coral 126
Plumarella sp. 126
poachers 63
Pododesmus macroschisma 182
Podothecus accipenserinus 64
polar eelpout 78
polar whelk 152
Polymastia sp. 108
Poroclinus rothrocki 82
predaceous acolis 178
Pribilof whelk 160
pricklebacks 80
Primnoa pacifica 126
Protothaca staminea 190
prowfish 88
Pseudarchaster parelii 262
Pseudarchaster parelii alascensis 262
Pseudopentaceros wheeleri 73
Psolus fabricii 300
Psolus japonicus 298
Psolus squamatus 300
Pteraster militaris 270
Pteraster obscurus 270
Pteraster octaster 270
Pteraster sp. 268
Pteraster tesselatus 268
Ptilosarcus gurneyi 130
Puget Sound king crab 220
Pugettia gracilis 236

purple hermit 244
purple-orange star 276
purple sea fan 124
purple soft coral 122
purple toothed henricia 274
Pycnopodia helianthoides 280
pygmy rock crab 240
Pyrulofusus deformis 162
Pyrulofusus harpa 162
Pyrulofusus melonis 162

Q

quillback rockfish 42

R

rainbow smelt 22
rainbow trout 16
Raja binoculata 8
Raja rhina 8
red-banded commensal worm 138
red cookie star 260
red frond-aeolis 176
red Irish lord 50
red king crab 226
red mushroom coral 120
red-orange armored cucumber 298
red salmon 20
red urchin 294
redbanded rockfish 38
redbanded star 282
reddish scallop 182
Reinhardtius evermanni (see *Atheresthes evermanni*)
Reinhardtius hippoglossoides 92
Reinhardtius stomias (see *Atheresthes stomias*)
remarkable sun star 264
rex sole 96
Rhamphocottus richardsonii 49
rhinoceros crab 222
Rhinoraja taranetzi (see *Bathyraja taranetzi*)
Rhinolithodes wosnessenskii 222
ribbed volute 168
righteye flounders 90
Rocinela angustata 208
rock greenling 46
rockfishes 27
ronquils 74
rose star 266
Rossia pacifica 202
rosy-undulate hydrocoral 118
rough China hat sponge 112
rough hydrocoral 118
rough keyhole limpet 144
rougheye rockfish 30
roughspine sculpin 50
rusty moonsnail 147

S

sablefish (blackcod) 44
saffron cod 26
sailfin sculpin 60
salmon 18
salmon snailfish 69
Salvelinus malma 18
sand dollar 296
sand lance 90
sandfish 88
Sarritor frenatus (see *Leptagonus frenatus*)
sawback poacher 63
Saxidomus gigantea 188
scaled crab 228
scallop sponge 114
scallops 182
scarlet king crab 228
scissortail sculpin 50
Sclerocrangon boreas 218
sculpins 49
sculptured shrimp 218
sea anemones 132
sea-clown triopha 174
sea cockroach 208
sea cucumbers 298
sea fan 122
sea football 300
sea mouse 136
sea onion 302
sea peach 304
sea pen 130
sea potato 304
sea raspberry 120
sea spider 204
sea stars 255
sea urchins 294
sea whip 130
searcher 74
Sebastes aleutianus 30
Sebastes alutus 32
Sebastes babcocki 38
Sebastes borealis 30
Sebastes brevispinis 42
Sebastes ciliatus 40
Sebastes flavidus 42
Sebastes maliger 42
Sebastes melanops 40
Sebastes melanostictus 32
Sebastes nebulosus 38
Sebastes nigrocinctus 38
Sebastes pinniger 36
Sebastes polyspinis 36
Sebastes ruberrimus 28
Sebastes variabilis 40
Sebastes variegatus 34
Sebastes zacentrus 34

Sebastolobus alascanus 28
segmented worms 136
serpent stars 286
serpent sun star 266
Serpula vermicularis 140
Serripes groenlandicus 186
Serripes laperousii 188
Serripes notabilis 188
shag-rug aeolis 178
sharks 4
sharpchin rockfish 34
shortfin eelpout 76
shorthorn sculpin 56
shortraker rockfish 30
shortspine thornyhead 28
shrew whelk 168
shrimps 211
sidestripe shrimp 212
Siliqua alta 196
Siliqua patula 196
silver salmon 18
silvergray rockfish 42
silverspotted sculpin 60
sinuous whelk 152
skates 7
skeleton shrimp 210
slippersnail 144
smelts 22
smooth leather bryozoan 252
smooth lumpsucker 66
snail eggs 170
snailfishes 68
snails 148
snake prickleback 80
snow crab 234
soccer ball sponge 104
sockeye (red) salmon 20
soft brown sponge 112
soft crab 224
Solaster sp. 264, 266
Solaster stimpsoni 264
Somniosus pacificus 4
southern rock sole 96
spectral snailfish 70
spiky arcturid 208
spiny arcturid 210
spiny-brown brittle star 288
spiny dogfish 6
spiny henricia 272
spiny king crab 226
spiny leather bryozoan 252
spiny lebbeid 216
spiny lithode crab 222
spiny paralomis 230
spiny spider crab 230
spinyhead sculpin 60
spiral bryozoan 250
splendid hermit 243
sponges 103
spot shrimp (prawn) 212
"spotted" diaulula 172
spud sponge 110
Squalus acanthias 6
squids 202
starry flounder 100
steelhead 17
Stegophiura ponderosa 292
stellate cookie star 258
stout eelblenny 82
striped sea leech 140
striped sun star 264
Strongylocentrotus droebachiensis 294
Strongylocentrotus fragilis 296
Strongylocentrotus franciscanus 294
Strongylocentrotus polyacanthus 294
sturgeon poacher 64
Styela rustica 304
Stylaster campylecus campylecus 116
Stylaster campylecus parageus 116
stylaster hydrocoral 116
Stylaster verrillii 116
Stylasterias forreri 282
Suberites domuncula 108
Suberites ficus 108
sunflower star 280
sunrise jellyfish 134
superb thouarella 128
sweet sea potato 302
Swiftia sp. 124
Synidotea sp. 208

T

Tanner crab 234
Tarasovium cornutum 204
Tellina lutea 194
Telmessus cheiragonus 238
Terebratalia transversa 200
tessellated slime star 268
Thaleichthys pacificus 24
Theragra chalcogramma 26
thick-cord whelk 154
Thouarella superba 128
threaded sculpin 53
threaded whelk 166
tiger rockfish 38
Tochuina tetraquetra 172
transverse lamp shell 200
tree sponge 108
Tresus capax 198
triangle Tanner crab 232
Trichodon trichodon 88
Trichotropis bicarinata 148
Triglops forficatus 50
Triglops macellus 50

Triopha catalinae 174
Tritonia sp. 174
tritonid (nudibranch) 174
trouts 16
tulip whelk 164
tumid henricia 272
tunicates 302
two-keel hairysnail 148

U

ubiquitous brittle star 288
Urticina crassicornis 132
Urticina felina (see *Urticina crassicornus*)
Utinomi's armored sea fan 128

V

variable skeleton sponges 104
variegated snailfish 68
vase sponge 103
Verrill's spider crab 230
"Vesteraalen" neptune 160
volute whelk 166
Volutopsius castanea 166
Volutopsius fragilis 164
Volutopsius middendorffii 164
Volutopsius trophonius 164
Vulcanella sp. 106

W

walleye pollock 26
warped whelk 162
Washington butterclam 188
wattled eelpout 76
weathervane scallop 184
white-scaled cucumber 300
white serpent star 287
white veiled chiton 142
whitebarred prickleback 82
whiteblotched skate 14
whitecap limpet 144
whiteknee hermit 248
widehand hermit 243
wolffish 86
wrinkled cushion star 270
wrymouths 84

Y

yellow Irish lord 52
yelloweye rockfish 28
yellowfin sole 100
yellowleg pandalid 216
yellowtail rockfish 42
yoldia 190
Yoldia sp. 190

Z

Zaprora silenus 88
Zirfaea pilsbryi 196

PICTURE INDEX

LAMPREY, page 4

SHARKS, page 4

SKATES, page 7

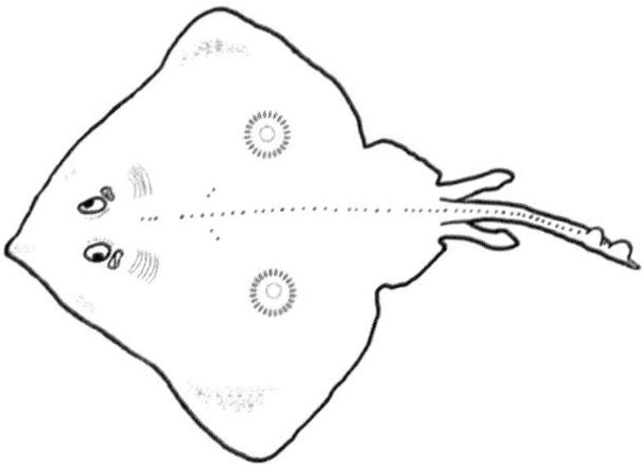

TROUTS, page 16

SALMON, page 18

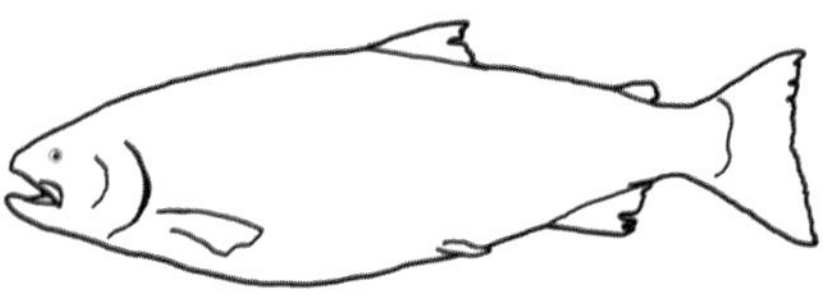

HERRING, page 22

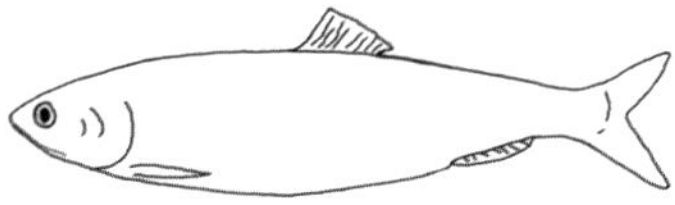

SMELTS, page 22

CODS, page 25

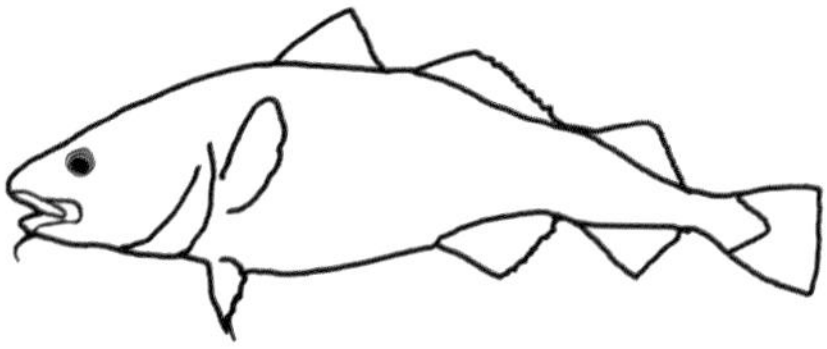

ROCKFISHES, page 27

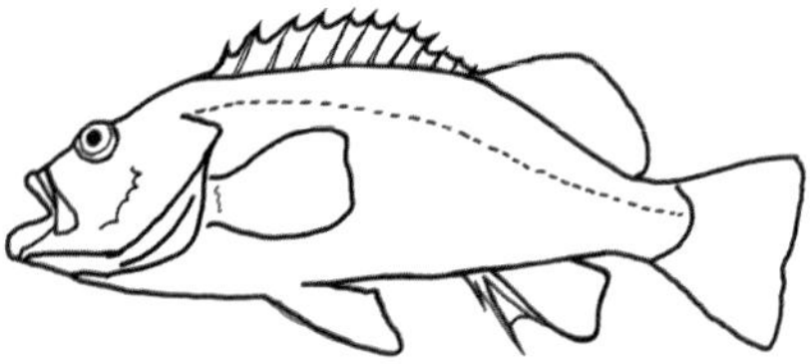

SABLEFISH, page 44

GREENLINGS, page 44

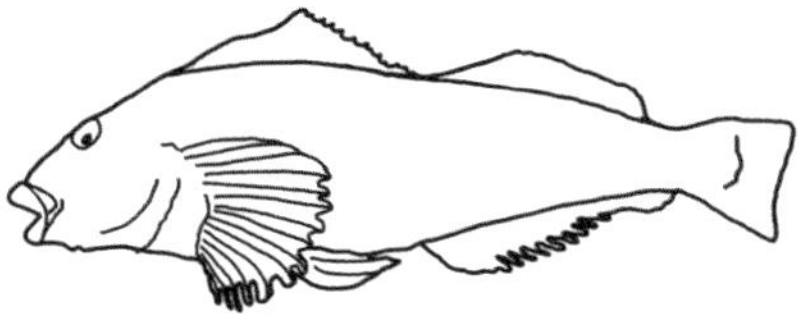

SCULPINS, page 49

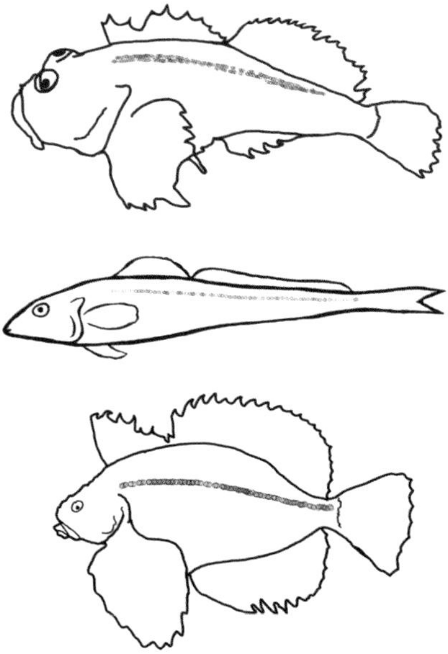

POACHERS, page 63

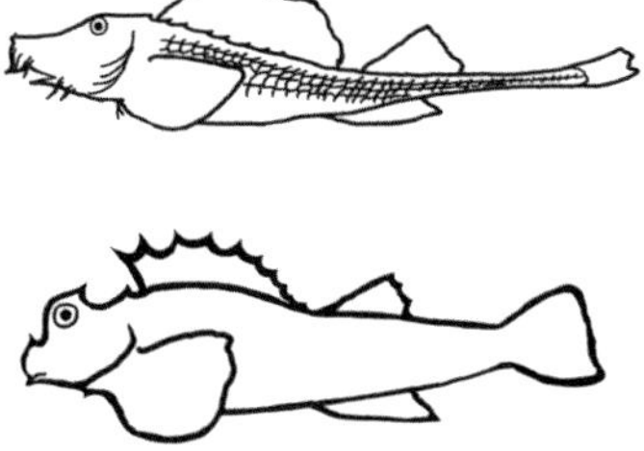

LUMPSUCKER, page 66

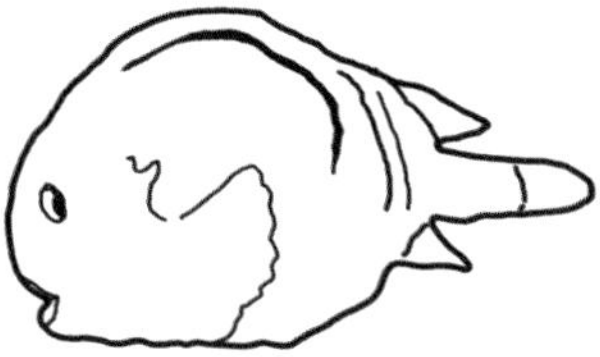

SNAILFISHES, page 68

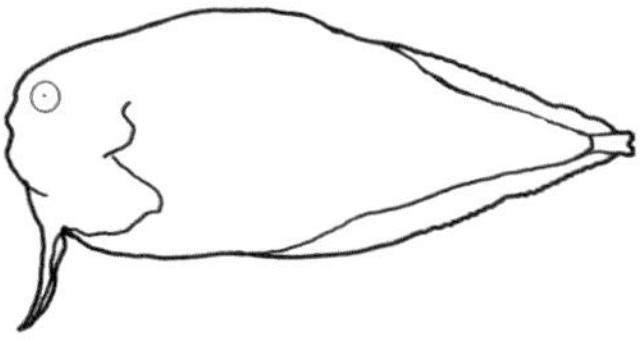

ARMORHEAD, page 73

RONQUILS, page 74

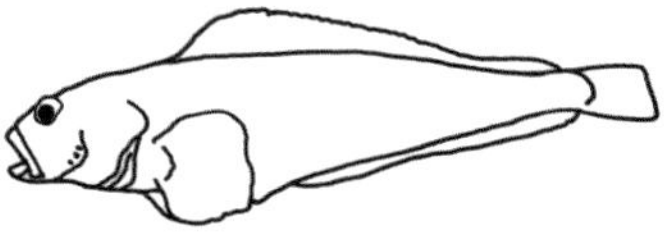

EELPOUTS, page 76

PRICKLEBACKS, page 80

WRYMOUTHS, page 84

GUNNEL, page 86

WOLFFISH, page 86

PROWFISH, page 88

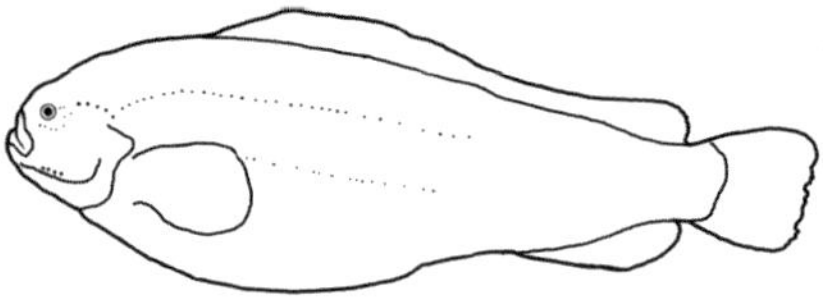

SANDFISH, page 88

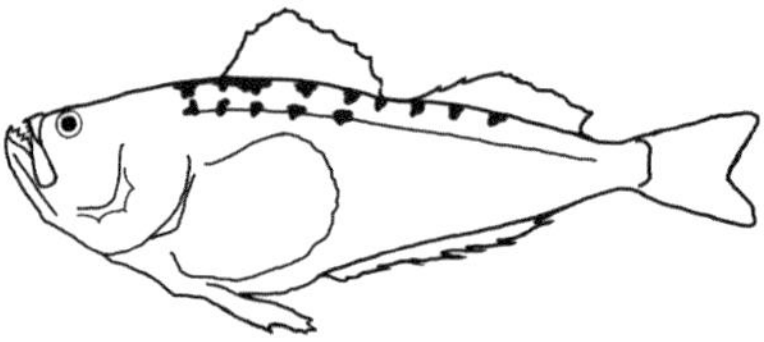

SAND LANCE, page 90

RIGHTEYE FLOUNDERS, page 90

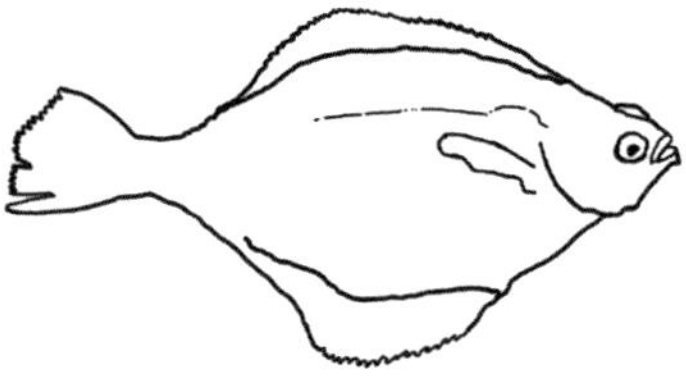

SPONGES, page 103

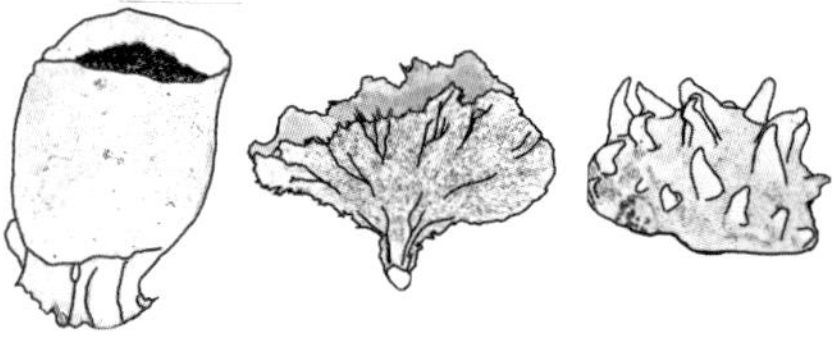

HYDROIDS, page 114

CORALS, page 116

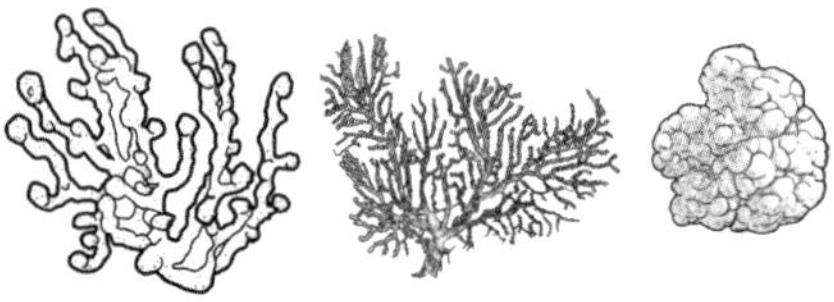

SEA PEN and SEA WHIP, page 130

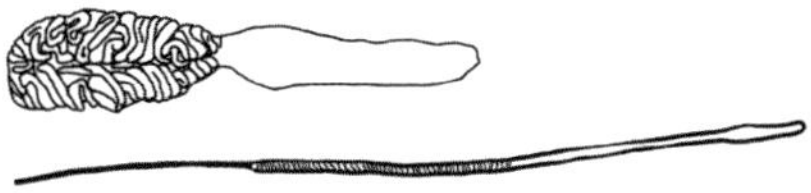

SEA ANEMONES, page 132

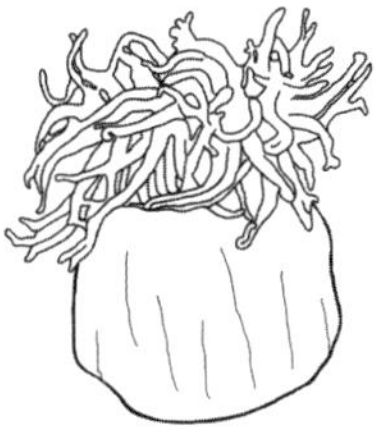

JELLYFISH, page 134

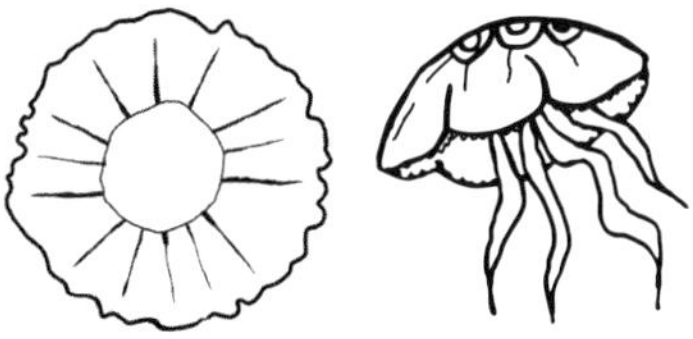

SEGMENTED WORMS, page 136

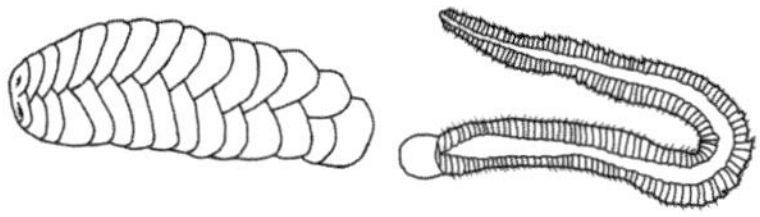

LEECH, page 140

CHITONS, page 140

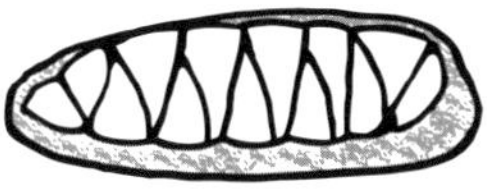

LIMPETS, page 144

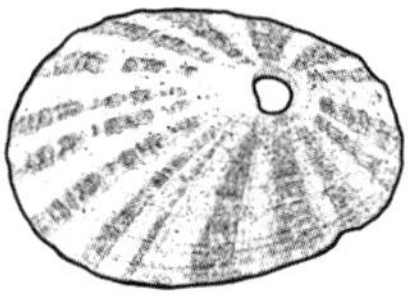

SLIPPERSNAIL, page 144

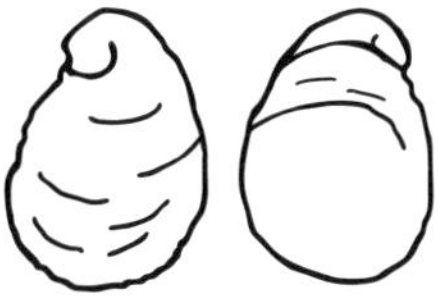

MOONSNAILS, page 147

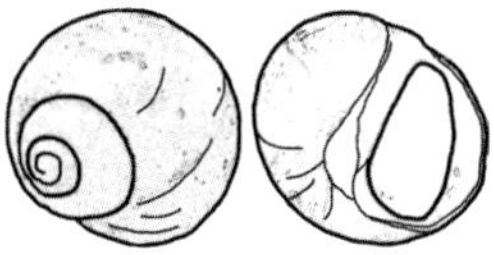

SNAILS, page 148

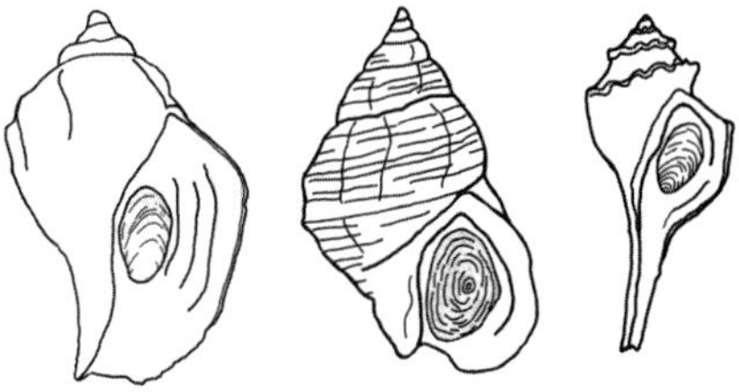

SNAIL EGGS, page 170

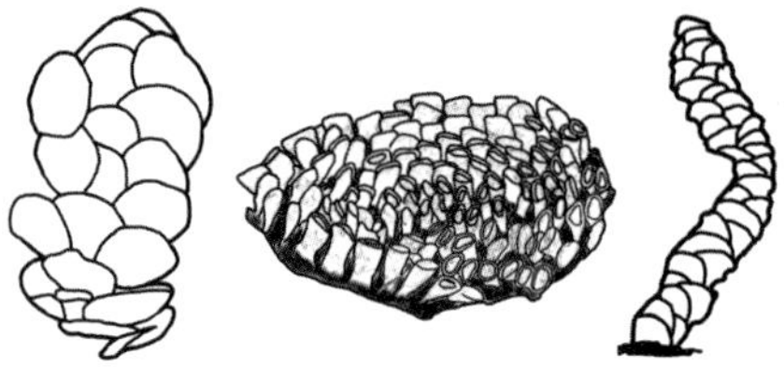

NUDIBRANCHS, page 172

MUSSELS, page 178

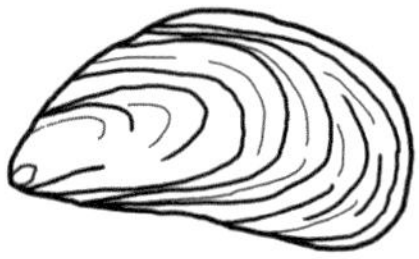

OYSTER, page 182

JINGLE, page 182

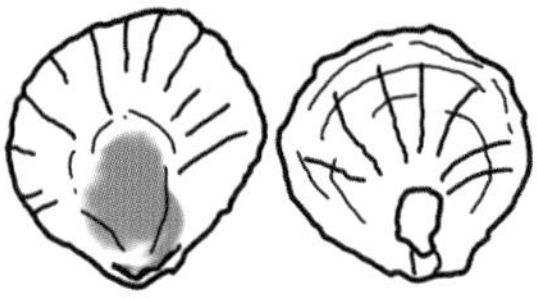

SCALLOPS, page 182

COCKLES, page 186

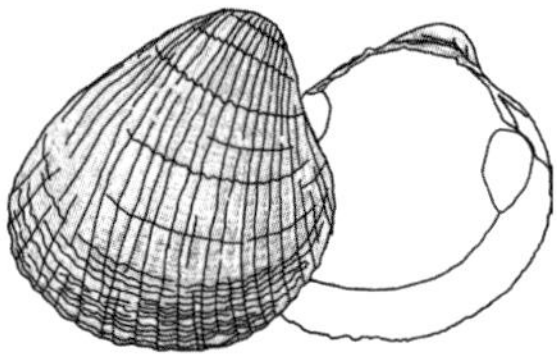

CLAMS, page 188

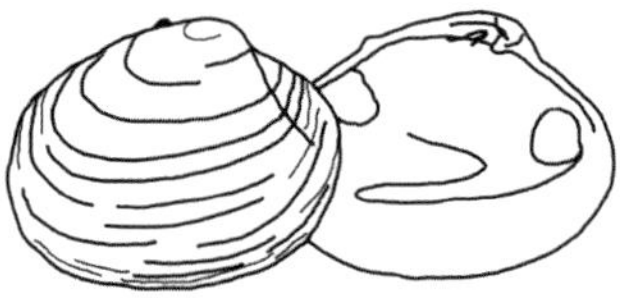

BRACHIOPODS, page 200

SQUIDS, page 202

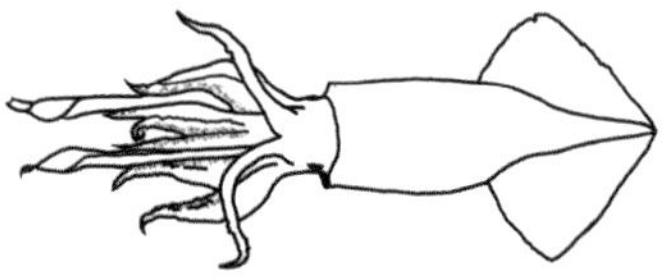

OCTOPUS, page 202

SEA SPIDER, page 204

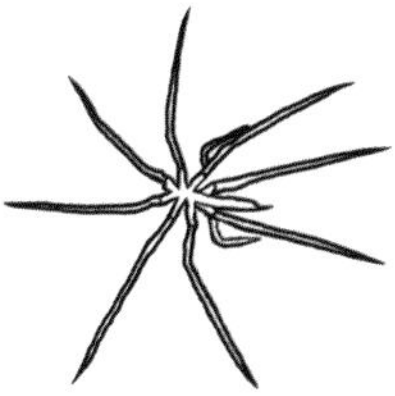

BARNACLES, page 204

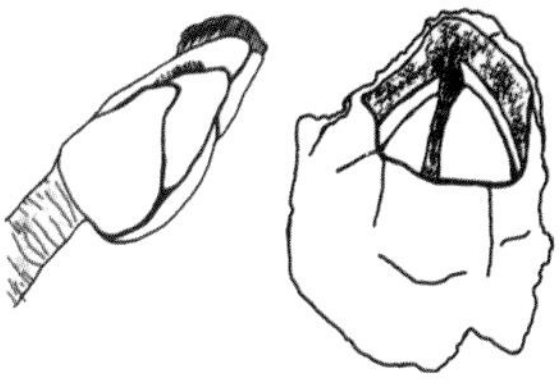

ISOPODS, page 208

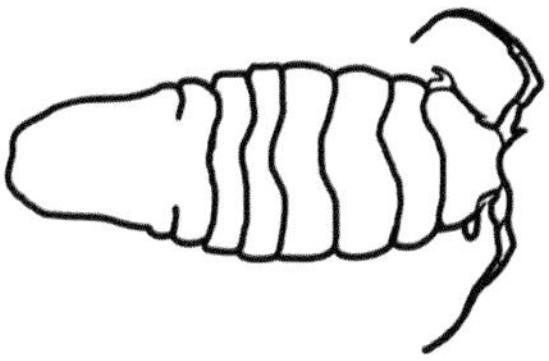

AMPHIPOD, page 210

SHRIMPS, page 211

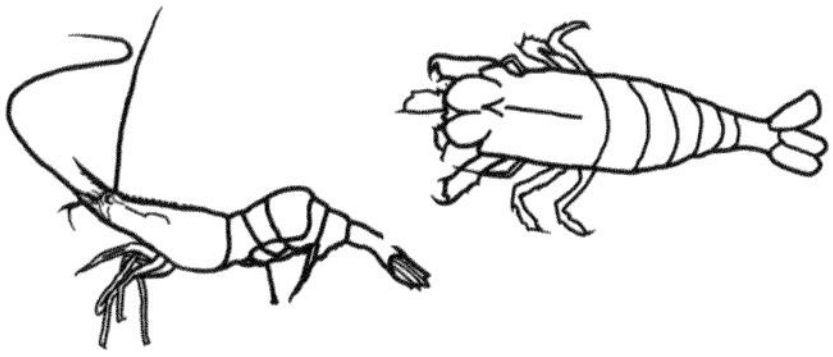

CRABS, page 220

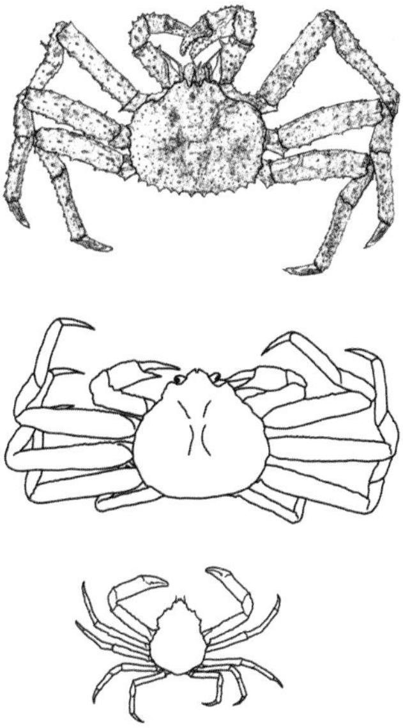

HERMIT CRABS, page 242

BRYOZOANS, page 250

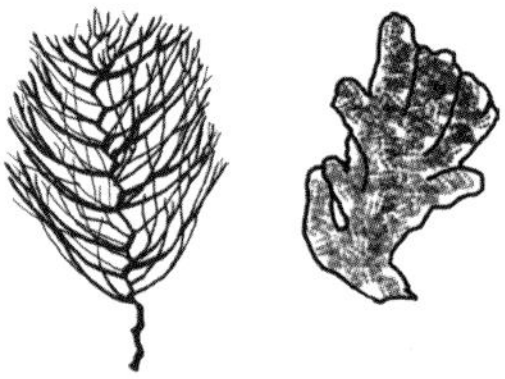

FEATHER STAR, page 254

SEA STARS, page 255

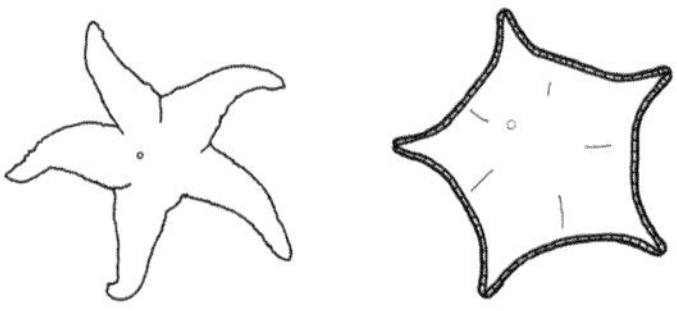

BASKET STAR, page 284

BRITTLESTARS, page 286

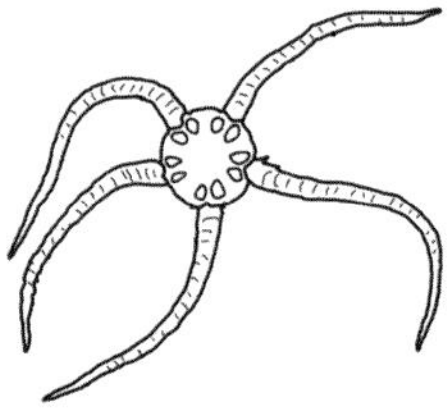

SEA URCHINS, page 294

SAND DOLLAR, page 296

SEA CUCUMBERS, page 298

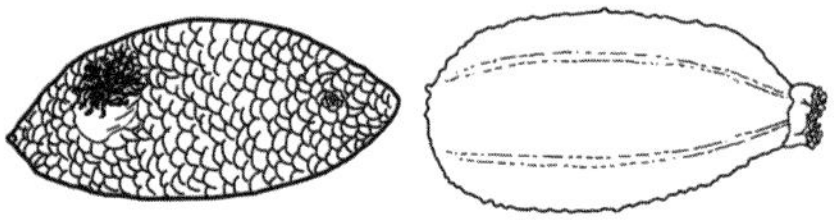

TUNICATES, page 302